Bioethanol Technologies

Edited by Freddie Inambao

Published in London, United Kingdom

IntechOpen

Supporting open minds since 2005

Bioethanol Technologies
http://dx.doi.org/10.5772/intechopen.91565
Edited by Freddie Inambao

Contributors
Prabakaran Balasubramanian, Jesús David Coral Medina, Antonio Irineudo Magalhaes Jr, Kristel M. Manzanero Gatdula, Butch G. Bataller, Rex B. Demafelis, Mohammad Moneruzzaman Khandaker, Umar Aliu Abdullah, Mahmoud Dogara Abdulrahman, Noor Afiza Badaluddin, Khamsah Suryati Mohd, Biancamaria Pietrangeli, Roberto Lauri, Octavio Armas, Reyes García-Contreras, Carmen Mata, María Cárdenas, Ifeanyichukwu Edeh

First published in London, United Kingdom, 2021 by IntechOpen
IntechOpen is the global imprint of INTECHOPEN LIMITED, registered in England and Wales, registration number: 11086078, 5 Princes Gate Court, London, SW7 2QJ, United Kingdom
Printed in Croatia

British Library Cataloguing-in-Publication Data
A catalogue record for this book is available from the British Library

Additional hard and PDF copies can be obtained from orders@intechopen.com

Bioethanol Technologies
Edited by Freddie Inambao
p. cm.
Print ISBN 978-1-83968-393-0
Online ISBN 978-1-83968-394-7
eBook (PDF) ISBN 978-1-83968-398-5

For EU product safety concerns:
IN TECH d.o.o., Prolaz Marije Krucifikse Kozulić 3, 51000 Rijeka, Croatia, info@intechopen.com or visit our website at intechopen.com.

Meet the editor

Freddie Inambao obtained his MSc in Mechanical Engineering in 1981 and Ph.D. in Technical Sciences in 1985. He is currently a professor at the University of KwaZulu-Natal, South Africa, and an advisor to the Green Energy Solutions Research Group at the same university. His expertise is in sustainable energy, energy management, energy efficiency, fuels, nanomaterials, and biomaterials. Dr. Inambao's research interests include energy efficiency of commercial and industrial buildings, piezoelectric materials (PZT) power generation, low-temperature solar thermal energy conversion, combustion of fossil fuels and renewable fuels, and desalination of water. Dr. Inambao has authored fifteen books with international publishers. He also has more than 160 research publications in peer-reviewed international journals to his credit. He was recognized as a top researcher at the University of KwaZulu-Natal in 2019.

Contents

Preface

Bioethanol is by far the most widely used biofuel for transportation worldwide. Bioethanol fuel has an important role in the field of environmental conservation by mitigating global warming, providing energy independence, offering new employment possibilities, and conserving fossil fuels. The demand for first-generation bioethanol (1G), produced mainly from agricultural crops, has continued to increase significantly over the past few years. These biofuels are derived mainly from edible food crops such as rice, wheat, barley, potato, corn, maize, sugarcane, starch-rich crops, and vegetable oil, for example, soybean oil, sunflower oil, olive oil, canola oil, mustard oil, and so on. Bioethanol is produced from these first-generation biofuels through fermentation. Second-generation bioethanol (2G) is the main feedstock for bioethanol production, which includes switchgrass, straw, cellulose, corn stover, and lignocellulosic biomass. Second-generation bioethanol are fuels that can be manufactured from various types of biomass. A major effort has begun to develop alternative ethanol feedstock using crop residues, forest by-products, perennial grasses, and other forms of plant biomass that are collectively termed "lignocellulosics." Lignocellulosic biomass can be broadly classified into virgin biomass, waste biomass (corn stover and straw included), and energy crops (switchgrass included). Second-generation bioethanol is produced from non-food crops such as wood, organic waste, food crop waste, and specific biomass crops. Second-generation bioethanol is more cost-competitive in comparison with existing fossil fuels. Third-generation bioethanol (3G) is the main feedstock for bioethanol production, represented by microalgae. Successful microalgae-production systems are the key to the development of the third bioethanol generation, as they could provide an alternative to one of the most difficult problems of large-scale biomass production deployment. Further, algae are the fastest-growing plants on earth. Improving the quality of air is one of the most important functions of bioethanol. When added to fuel, bioethanol reduces the use of cancer-causing gasoline compounds such as ethylbenzene, xylene, toluene, and benzene. It also reduces the emissions of small particulates and soot from motor fuels as well as greenhouse gas emissions. Fourth-generation (4G) bioethanol is obtained from the modification of Escherichia coli gene altercations through the application of metabolic engineering or systems biology strategies. The number of jobs created directly from advanced biofuels production is expected to grow in the future. Green jobs are activities characterized by previously evaluated risks, but with a different scope and exposition in connection with newly applied technology. Therefore, it is strategic and important to complete the risk assessment process with respect to new or emergent risks. An inclusive sustainability assessment of bioethanol production alternatives should incorporate an occupational health and safety (OH&S) assessment, as it is necessary to integrate health and safety issues at the early stages of development of the industrial process in order to define tailored mitigation measures at full-scale plants.

This book contains seven chapters. Chapter 1 covers the state-of-the-art processes involved in bioethanol production including pretreatment, hydrolysis, fermentation processes, bioethanol recovery, integrated processes, life cycle assessment, techno-economic analysis, exergy analysis, and process simulation. Chapter 2 presents an overview of second-generation ethanol (2GE) production and the possibilities of co-production of high value-added molecules and their economic and environmental assessment, including CO2 release, water consumption, solid residue disposal, and economic analysis to determine the best bioethanol-based biorefinery configuration. Chapter 3 provides a review of the procedures involved in the production of bioethanol from biomass of fruits and vegetable waste through a fermentation process using Saccharomyces cerevisiae. In this chapter, the authors discuss biomass preparation and fermentation techniques for bioethanol and review the results of different fruits and vegetable waste. Chapter 4 consolidates information on the potential of different biomass available in the Philippines that can be utilized as feedstock for bioethanol production. It presents a brief overview of the bioethanol market in the Philippines, which supports the need for complementary feedstock to sugarcane for the different generations of bioethanol. It also provides the criteria for selecting good feedstock for bioethanol, examines challenges that may be encountered upon using the biomass as starting raw material, and discusses appropriate conversion technologies. Chapter 5 provides information on blending higher volumes of ethanol with diesel for replacing the neat diesel to fuel compression ignition engines. Chapter 6 presents different experimental works that study the effect of bioethanol-diesel fuel blends and their effects on the integrity of some parts of a diesel injection system and on performance and regulated emissions of engines and/or vehicles under different transient conditions. The studies described are carried out in an engine test bench, two public buses under urban transportation, and a construction machine in actual railway construction. In biofuel production plants, some work activities in processing the biomass are sources of airborne dust and thus employers should demonstrate that adequate control measures have been implemented to prevent workers' exposure. Chapter 7 analyzes the production process of a 2G bioethanol plant in order to specify the process phases that cause occupational health issues related to airborne dust. The chapter also provides technical recommendations for combatting this type of occupational hazard.

Sharing the book with readers worldwide, I am grateful to a number of individuals who have contributed to this book. In particular, I would like to thank all the authors for their contributions and the publishing staff at IntechOpen. I am especially very grateful to Author Service Manager Ms. Dolores Kuzelj for making sincere efforts for the book's timely completion.

Freddie Inambao
Professor,
Howard College,
Faculty of Engineering,
Mechanical Engineering,
University of KwaZulu-Natal,
Durban, South Africa

Chapter 1

Bioethanol Production: An Overview

Ifeanyichukwu Edeh

Abstract

Bioethanol is currently being considered as a potential replacement for the conventional gasoline, especially as it possesses similar and some superior qualities enabling reduction in GHG and increases fuel reserve. Bioethanol used for commercial purposes is usually produced from edible feedstocks such as corn and sugar cane which increases the production cost. The high cost of these feedstocks is the driving force behind the search for the second, and third generations (3G) bioethanol produced from cheaper and available feedstocks. The fourth-generation bioethanol is being developed to further advance the 3G bioethanol to enhance the potential of algae to capture CO_2 and to increase the production of specific compounds. Despite the efforts been made to reduce the cost of production through the use of diverse non-edible feedstocks, the cost of processing the feedstocks is still very high, thereby making bioethanol uncompetitive with the conventional gasoline. The life cycle assessment and techno-economic analyses are usually conducted to assess the economic feasibility and the environmental impact of the bioethanol production processes. This chapter thus, covers the State-of-the-art processes involved in bioethanol production including pretreatment, hydrolysis, fermentation processes, bioethanol recovery, integrated processes, Life cycle assessment, techno-economic analysis, exergy analysis and process simulation.

Keywords: bioethanol, lignocellulose, hydrolysis, pretreatment, fermentation, distillation, exergy, simulation, techno-economic, life cycle assessment

1. Introduction

The depletion of the fossil fuel and global warming caused by the emission of greenhouse gases from the combustion of fossil is currently driving researchers in the direction of finding alternative and environmentally friendly fuel. Biofuels are one of the numerous options being considered. Bioethanol is considered as the most promising biofuel to replace gasoline, especially due to its properties. This biofuel is a liquid oxygenated fuel containing 35% oxygen produced from the microbial fermentation of monomeric sugar obtained from carbohydrate sources such as corn, soybeans and sugar cane. The bioethanol produced globally in 2018 was 110 billion liters and is expected to increase to 140 billion liters in 2022 with compound annual growth rate (CAGR) of 7.6% due to anticipated economic feasibility of the process [1]. The US, Brazil, European Union, China and Canada respectively are the global powerhouses in bioethanol production. The US uses corn as the feedstock to produce bioethanol and obtained a production capacity of ~57.7 billion liters while

Brazil produces bioethanol from sugar cane and had a total production capacity of ~27.6 billion liters in 2016 respectively [2].

Bioethanol is considered a potential substitute for the conventional gasoline and can be used directly in vehicles or blended with the gasoline, thereby reducing greenhouse gas emissions and consumption of gasoline [3]. For direct application (E100), the timing (and electronic control system if in use) of the gasoline engine is adjusted, and larger gasoline tank is used. However, the use of bioethanol (E100) is usually characterized with difficulty in starting the engine at a low temperature or during the cold weather due to higher heat of vaporization. Required. The blending of bioethanol with gasoline might not require modifying the engine, rather it will help to enhance ignition or engine performance. The most commonly used blends are E85 and E10. Advantages of bioethanol include high-octane rating resulting to increased engine efficiency and performance, low boiling point, broad flammability, higher compression ratio and heat of vaporization, comparable energy content, reduced burning time and lean burn engine [4]. The disadvantages include high production cost resulting from high cost of feedstock, enzymes, detoxification and ethanol recovery, respectively. Bioethanol possesses a low volumetric energy density, meaning that more volume of bioethanol/km (up 50%) will be consume compared to the conventional gasoline [3]. The use of bioethanol in engines might require frequent replacing the engine parts as the bioethanol has the capacity to degrade some elastomers and cause corrosion of metals [5]. However, in attempt to reduce the cost of production, lignocellulosic biomass is being considered as feedstocks because of availability and low cost of acquisition. Unfortunately, the processing cost is still high, thereby, making the process unattractive economically [6].

When bioethanol is produced from edible feedstocks such as corn and sugar cane, it is called first generation (1G) bioethanol and 2G second-generation (2G) bioethanol if the feedstock is a lignocellulose. Examples of these lignocellulose biomass is switch grass, cornstalks, wood, herbaceous crops, waste paper and paper products, agricultural and forestry residues, pulp and paper mill waste, municipal solid waste and food industry waste. Lignocellulosic biomass is made up of cellulose, hemicellulose, lignin, protein, ash, and minor extractives [7]. Lignocellulosic biomass is being considered as feedstocks for bioethanol production due to relatively low cost of acquisition, availability and sustainability of supply. This biomass has the capacity to increase the current production rate of bioethanol and is being speculated to produce approximately 442 billion liters per year of bioethanol globally. The 2G-bioethanol has a greater potential to reduce the greenhouse gases emission compared to 1G -bioethanol. The third generation (3G) bioethanol is obtained when algae are used as the feedstock. Algae bioethanol is gaining traction possibly due to high carbohydrate content and absence of lignin in most available algae. With this kind of feedstock, the cost of pretreatment is expected to reduce as the complex

Algae	Bioethanol yield (%)	Ref.
Nannochloropsis Oculata	3.68	[9]
Tetraselmis suecica	7.26	[9]
Scenedesmus dimorphus	49.7	[10]
Porphyridium cruemtum (seawater)	65.4	[11]
Porphyridium cruemtum (fresh water)	70.3	[12]
Padina Tetrastromatica	16.1	[12]

Table 1.
Yield of difference species of algae.

lignin removal process is eliminated [8]. Numerous researchers have investigated the use of algae as feedstock for bioethanol production. Based on the results obtained, the species of algae with high productivity are presented in **Table 1**.

The fourth-generation (4G) bioethanol is obtained from the modification of *E. coli* gene altercations through the application of metabolic engineering or systems biology strategies [13].

2. Bioethanol production process

The processes involved in the production of bioethanol from different feedstocks include pretreatment, hydrolysis, fermentation and ethanol recovery. These processes are explained below:

2.1 Pretreatment

Pretreatment is one of the costliest steps in the production of bioethanol from lignocellulose biomass accounting to approximately $0.30/gallon of ethanol produced. There exist different pretreatment methods aimed at increasing the reactivity of cellulose and the potential yield of the fermentable sugars. These may be either traditional or advanced pretreatments. Traditional pretreatments are classified into four categories which include chemical, physical, physicochemical, and biological methods while advanced pretreatment method may be either acid-based fractionation or ionic liquid-based fractionation (ILF) [14]. Amongst the traditional pretreatment methods, chemical categories are the most efficient and hence predominantly used [15].

The pretreatment of lignocellulosic biomass through various methods helps to release cellulose usually embedded in a matrix of polymers consisting of lignin and hemicellulose by disrupting the original structure (**Figure 1**). With this, cellulose is separated from the polymer matrix and is more accessible for enzymatic hydrolysis, thereby resulting to increased sugar yields greater than 90% (theoretical yield) using feedstocks such as grasses, corn and wood [16]. This means that cellulose is more susceptible to enzymatic hydrolysis when its crystalline structure is disrupted. Without the disruption, enzymes bind on the surface of the lignin and not the cellulose chains impeding enzymatic hydrolysis.

Other advantages of pretreatment include helping to prevent the degradation of sugars (pentoses); ensuring viability of the bioethanol production processes by using moderate size reactors and minimizing heat and power requirements, and minimizing the formation of inhibitors which reduces the yield of the hydrolysis and hence the fermentation of sugar to ethanol [16].

2.1.1 Traditional pretreatments

These pretreatments method have been discussed extensively in the literature. As mentioned earlier, the method is categorized as: (1) physical pretreatment- this involves the breaking down of the size of the lignocellulosic biomass and crystallinity by methods such as milling, grinding, irradiation and extrusion. The resultant effect of which are increased surfaced area and pore size of the biomass enabling increase in the enzymatic hydrolysis efficiency. Physical pretreatment may need combining with chemical pretreatment to enhance the efficiency of deconstruction of the lignocellulose [17]. (2) Chemical pretreatment: these include acid, alkali, oxidative delignification, and organic acid (organosolvation) methods. They are highly selective for specific type of feedstocks, and are used to deconstruct and remove lignin and/or hemicellulose from the polymer matrix. Chemical pretreatments

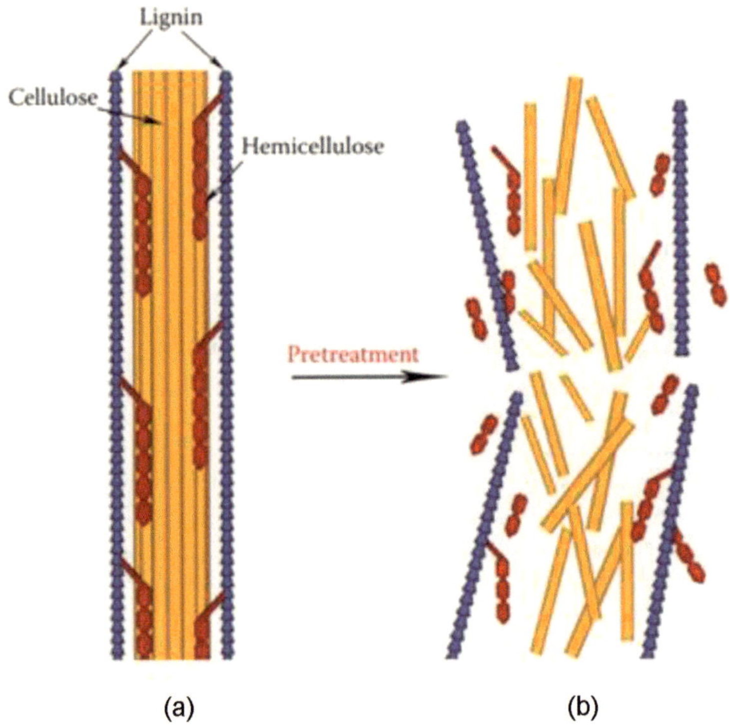

Figure 1.
Effect of pretreatment on the lignocellulosic biomass [16]. (a) Lignocellulosic biomass before pretreatment, and (b) Lignocellulosic biomass after pretreatment.

are undoubtedly effective but require harsh operating conditions which may have adverse effect on the downstream processing and the by-products may need special disposal procedures [17]. (3) Physicochemical: this combines the features of both physical and chemical pretreatments. Examples are steam explosion, liquid hot water, microwave irradiation and CO_2 explosion [18]. (4) Biological pretreatment: This involves the use of microorganisms to breakdown lignocellulosic biomass for further enzymatic hydrolysis. These organisms include white-rot, brown-rot and soft-rot fungi, and bacteria [19].

2.1.2 Advanced pretreatment methods for lignocellulose

These methods are also called lignocellulose fractionation pretreatment and are targeted at reducing the cost of cellulosic ethanol production by fractionating the lignocellulose in such a way to generate value-added co-products under a mild operating condition like 50°C and atmospheric pressure [20]. This gain is achieved by using cellulose solvents which enhances the cellulose accessibility and separation of cellulose, hemicellulose, and lignin to produce value-added co-products [21]. The method is also known as Cellulose solvent-based lignocellulose fractionation (CSLF). The operation helps to reduce the quantities of enzymes required for the subsequent enzymatic hydrolysis and could be used for varieties of feedstocks [21].

There are two general techniques used in CSLF which include (1) acid-mediated fractionation and (2). Ionic liquid-based fractionation (ILF). These are discussed below:

2.1.2.1 Acid-mediated fractionation

The cellulose solvents such as phospholic acid and organic solvents like acetone or ethanol are usually used at mild operating conditions of 1 atm and 50°C to separate lignocellulosic biomass. The effectiveness of the separation is dependent on the solubility properties of the cellulose, hemicellulose, and lignin in the cellulose solvent, organic solvent and water, respectively [20]. Separating lignin and hemicellulose from the cellulose fraction helps to reduce substrate recalcitrant and competitive binding sites, unwanted sugar degradation, cost and production of the inhibitors [20]. This method has been used efficiently to pretreat varieties of lignocellulose such as bamboo, corn stover, sugarcane, switchgrass and elephant grass [22].

2.1.2.2 Ionic liquid-based fractionation

Ionic liquids (ILs) are salt solutions consisting of significant quantity of organic cations and small/inorganic anions that exists as liquid at relatively low temperatures like room temperature. They are used to fractionate lignocellulose to obtain specific, purified and polymeric raw materials which are intact and are easily separated and used as value-added co-products. In comparison of the conventional lignocellulosic biomass, ILs pretreatment methods show some advantages such as less energy intensive, simplicity of operation and capacity to separate specific components [20]. The properties such as low vapor pressure and high thermal stability suggest that ILs are environmentally friendly and as such are considered as green solvents. ILs are also considered to be tunable due to such properties as hydrophobicity, polarity, and solvent power which can be adjusted to achieve specific desirable results. These properties of ILs with those of antisolvent and lignocellulose (type, moisture content, partial size, and load) with temperature, pretreatment time can be used to determine the overall efficiency of the ionic liquid pretreatment method [23].

However, the most frequently used pretreatment method is steam explosion. This patronage could be due to its low capital investment, high energy efficiency, less environmental impact, less hazardous process chemicals and conditions, and complete sugar recovery [24].

2.2 Hydrolysis

Following the pretreatment of the lignocellulosic biomass is the hydrolysis of polymeric carbohydrate (cellulose and hemicellulose) to produce sugar monomers. This stage is required since enzymes needed in the succeeding stage (fermentation) can only digest sugar monomers. The process can be catalyzed either by acid or enzymes. Acid-catalyzed hydrolysis is the most commonly used method and it involves either the use of concentrated or dilute acid (see Eq. (1)). Example of such acids are H_2SO_4 and HCl. The concentrated acid-catalyzed hydrolysis is used at lower temperature and high acid concentration, resulting to 90% sugar recovery at a short period of time [25]. The disadvantage of this method is the high cost of production due to difficulty in acid recovery, disposal, concentration control and recycling [26]. Another problem with the concentrated acid-catalyzed hydrolysis treatment is its capability to degrade sugar monomers due to the prevailing acidic environment. The dilute acid-catalyzed hydrolysis requires high temperature and low acid concentration. The most predominantly used acid is dilute acid. The problem with this method of hydrolysis is that the process results to the formation of inhibitors compared to the acid-catalyzed hydrolysis.

$$\left(C_6H_{10}O_5\right)_n + H_3O^+ \rightarrow H_2O + nC_6H_{12}O_6 \quad (1)$$

Acid hydrolysis of the lignocellulose is carried out in two stages. Stage one is where the hemicellulose is hydrolyzed with the help of dilute acid and in the second stage, cellulose is hydrolyzed using concentrated acid [25].

Enzyme-catalyzed hydrolysis uses enzymes to hydrolyze polymeric carbohydrate to sugar monomers under mild operating conditions of temperature 45–50°C and pH 4.8–5.0. This method is efficient and results to high sugar recovery without inhibitor formation and tendency to cause corrosion. The efficacy of the enzyme-catalyzed hydrolysis is affected by factors such as pH, enzyme loading, time, temperature and substrate concentration. The hydrolytic process can be catalyzed by three kinds of cellulase enzymes, name endo-1,4-β-glucanases, cellobiohydrolases and β-glucosidases. These enzymes are usually very expensive due to high demand from various industries such as paper, textile and food processing industries [1]. The high cost of these enzymes also impacts on the overall cost of production especially as large quantities of enzymes are required. Based on the cost, microorganisms with the potential of secreting cellulolytic enzymes are broadly used in the contemporary times. These include *Clostridium*, *cellulomonas*, *Erwinia, Thermonospora, Bacteriodes, Bacillus, Ruminococcus, Acetovibrio*, and *Streptomyces.* Others include fungi such as *Trichoderma, Penicillium, Fusarium, Phanerochaete, Humicola, and Schizophillum sp*. The most commonly used microbial enzymes amongst these microorganisms is Trichoderma species [27]. The problems with the microbial enzymes are stability, substrate or product inhibition and catalytic efficiency. Although, with advances in genetic modifications, recombinant DNA techniques and application of various strategies to improve the strains help to increase the quantity of enzymes produced, make them more robust and economically feasible. The efficiency of the cellulose hydrolysis can also be improved by the addition of Polyethylene glycol (PEG) or Tween 20 resulting to increased enzymatic saccharification and reduction in the adsorption of cellulose on lignin [25].

The mechanism of the hydrolysis of lignocellulosic biomass to glucose occurs in three steps and are presented in **Figures 2–6**. The first step is the linking of the β-1,4 bond of the cellulose with water molecule catalyzed by endoglucanase (1,4-β-D-glucanohydrolase) resulting to the formation of cellodextrin with a shorter chain, and free-chains ends (reducing and non-reducing ends) (**Figure 2**) [28]. The second step is the degrading of cellodextrin to a two-unit glucoses (cellobioses) with the help of exoglucanase (1,4- β-D-glucan cellobiohydrolase) by adjusting the reducing and non-reducing chains (**Figure 3**) [29]. The third step is the formation of glucose obtained when the β-glucosidases strikes the cellobioses (**Figure 3**) [30]. The production of glucose is necessary because, the subsequent process which is fermentation requires the use of the simplest monomer as feedstock.

The hydrolysis of hemicellulose is easier compared to cellulose due to its possession of more amorphous property. The hemicellulose contains 10–15% and 10–35% of xylan in soft and hard woods, respectively. Xylan has both main and outer chains. The former can be degraded using endo-β-1,4-xylanase (EC 3.2.1.8) and β-xylosidase (EC 3.2.1.37). The main chain of xylan is hydrolyzed to a short chain xylan oligosaccharide through the help of endo-β-1,4-xylanase (**Figure 5**). The oligosaccharide is further degraded to a pyranose form of xylan known as xyropyranose by β -xylosidase (**Figure 6**) [32]. On the contrary, the outer chains of the xylan

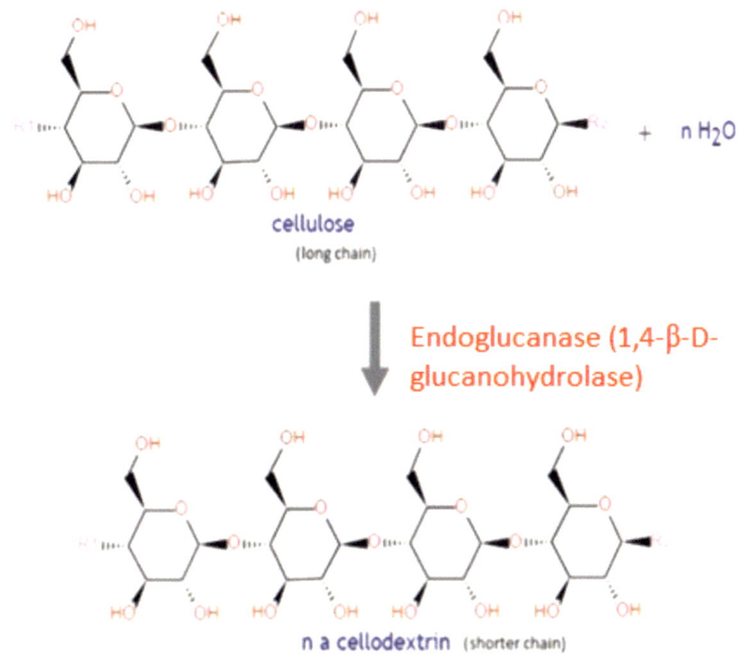

Figure 2.
Hydrolysis of long chain cellulose to a shorter chain cellulose (cellodextrin) [28].

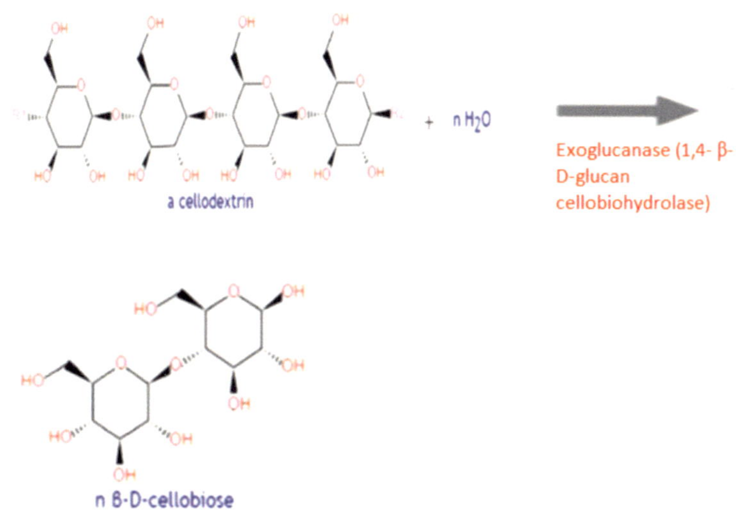

Figure 3.
Hydrolysis of cellodextrin to cellobiose catalyzed by exoglucanase (1,4- β-D-glucan cellobiohydrolase) [29].

can be degraded by enzymes known as accessory xylanolytic enzymes such as feruloyl esterase (EC 3.1.1.73), α-L-arabinofuranosidase (EC 3.2.1.55), α-glucuronidase (EC 3.2.1.139), and acetylxylan esterase (EC 3.1.1.72).

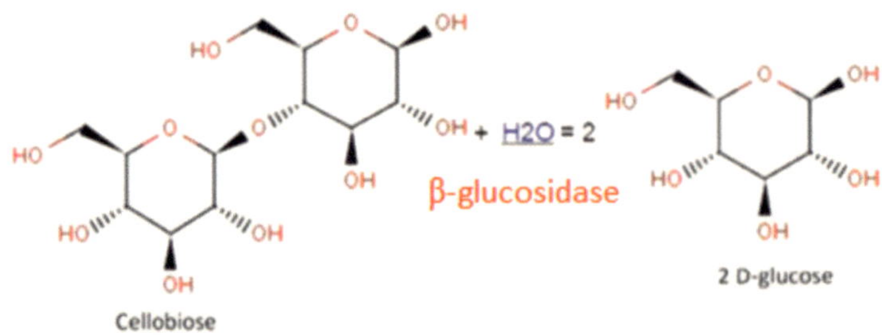

Figure 4.
Hydrolysis of cellobiose to 2 D-glucose catalyzed by β-glucosidase [30].

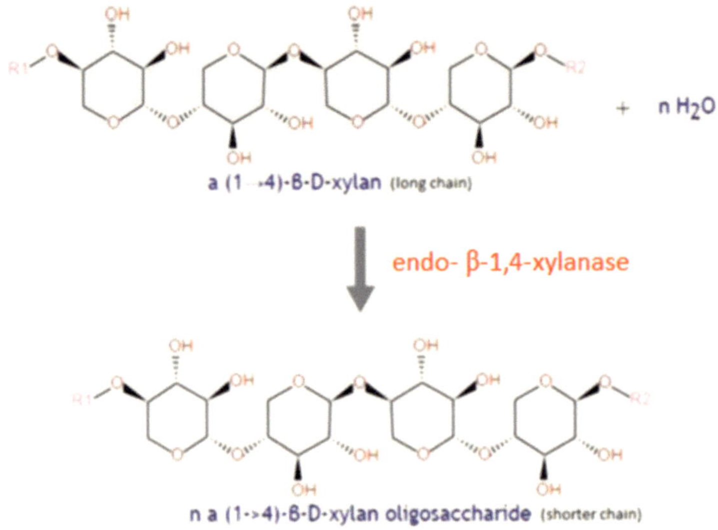

Figure 5.
Hydrolysis of long chain xylan to a shorter chain xylan oligosaccharides by endo-β-1,4-xylanase [31].

2.3 Fermentation processes

This is a biological process that involves the conversion of the monomeric units of sugars obtained from the hydrolysis step into ethanol, acids and gases using microorganisms such as yeast, fungi or bacteria (see Eq. (2)) [1, 33]. The most commonly used microorganism is yeast especially *Saccharomyces cerevisiae* due to high yield of ethanol and high tolerance limits [34]. *Saccharomyces cerevisiae* converts glucose,

$$C_6H_{12}O_6 + yeast \rightarrow 2C_2H_5OH + 2CO_2 \tag{2}$$

mannose or fructose which can be obtained from the hydrolysis of cellulose to ethanol while xylan from the hydrolysis of hemicellulose can be converted to xylose. Some examples of different microorganisms used in fermentation of simple sugars and their respective ethanol yields at varying operating conditions are presented in **Table 2**. As can be seen, *Saccharomyces Cerevisiae 3013* followed by *Zymomonas Mobilis ZMA7-2* gave the maximum ethanol yield [33].

2.3.1 Fermentation technologies

The technologies used for the fermentation of monomeric units of sugar to ethanol include separate hydrolysis and fermentation, simultaneous

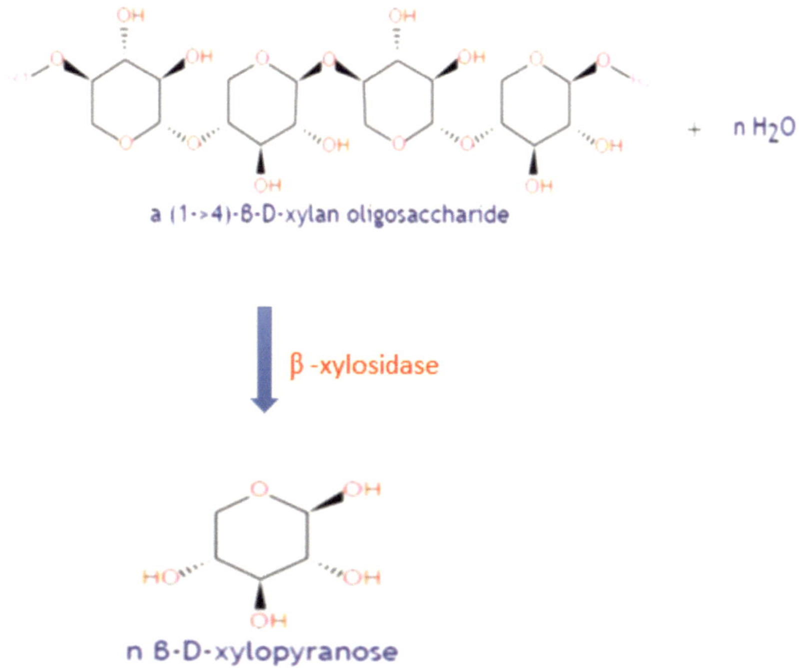

Figure 6.
Hydrolysis of xylan oligosaccharide to xylopyranose by β-xylosidase [32].

saccharification and fermentation (SSF), simultaneous saccharification and co-fermentation (SSCF), non-isothermal simultaneous saccharification and fermentation, simultaneous saccharification, filtration and fermentation, consolidated bioprocessing (CBP). The first three are commonly used [26]. Other types of fermentation include batch, fed-batch, continuous and solid-state fermentation. Some of these fermentation methods are discussed below.

2.3.1.1 Batch fermentation process

This is the simplest of the fermentation processes as it is flexible for a range of products, easy to control and has multi-vessel. The process involves adding the substrates, microorganism, culture medium and nutrients at the beginning of the operation in a closed system under favourable conditions at a predetermined time. The products are only withdrawn at the end of the fermentation time. The problems with this type of fermentation process are low yield, long fermentation time, and high labour cost making batch process unattractive for commercial production of bioethanol [26, 40]. Also, due to high sugar concentration in the fermentation medium, there could be substrate inhibition leading to inhibition of cell growth and ethanol production [41].

2.3.1.2 Continuous fermentation process

This process involves adding substrates, culture medium and nutrients into a fermentor containing active microorganisms and withdrawing the products continuously. The products obtained are usually ethanol, cells and residual sugar. The advantages of continuous fermentation process are high productivity, small fermenter volumes, and low investment and operational cost [42]. The disadvantages include possibility of product contamination, and potential decline in yeast capability to support ethanol production because of long cultivation time [43].

Microorganism		Temperature (°C)	pH	Fermentation time (h)	Sugar concentration (g/L)	Ethanol yield (g/L)	Ref.
Yeasts	*Saccharomyces Cerevisiae 3013*	30	5.5	65	280	130.12	[35]
	Saccharomyces Cerevisiae BY4742	35	5.0	96	80	39	[36]
Bacteria	*Zymomonas Mobilis NRRL 806*	30	6.5	18	117	30.4	[37]
	Zymomonas Mobilis ZMA7-2	30	4.0	44	200	99.78	[38]
Fungi	*Aspergillus oryzae 694*	First aerobic step (30)	5.0	24	50	24.4	[39]
		Second anaerobic step (30)	5.0	144			
	Rhizobium javanicus 2871	First aerobic step (30)	5.0	24	100	33	[39]
		Second anaerobic step (30)	5.0	72			

Table 2.
Effects of microorganisms on the yield of ethanol under varying operating conditions [33].

2.3.1.3 Fed-batch fermentation process

This is the combination of batch and continuous fermentation processes involving charging the substrate into the fermentor without removing the medium. Comparing with other fermentation processes, fed-batch process has higher productivity, more dissolved oxygen in medium, shorter fermentation time and lower toxic effect of the medium [43]. The disadvantage is that ethanol productivity is limited by cell mass concentration and feed rate [40].

2.3.1.4 Separate hydrolysis and fermentation (SHF)

The Enzymatic hydrolysis is separated from fermentation allowing enzymes to operate at high temperature and the fermentation microorganisms to function at moderate temperature for optimum performance [26]. Since the hydrolytic enzymes and the fermentation organisms operate at their optimum conditions, it is expected that the productivity of ethanol will be high. The disadvantages of SHF are high capital cost especially as two reactors are required, requirement of high reaction time, and possibility of limiting the cellulase activities by sugars released during the hydrolysis step [44].

2.3.1.5 Simultaneous saccharification and fermentation (SSF)

Here the saccharification of cellulose and the fermentation of monomeric sugars are carried out in the same reactor simultaneously [45]. Since the hydrolysate is simultaneously used for fermentation, the usual inhibition of the cellulase activities can be avoided [46]. The disadvantage of SSF is the variation in the optimum temperature required for efficient performance of the cellulase and microorganisms during hydrolysis and fermentation, respectively. The high temperature required by the cellulase for hydrolysis might reduce the microorganisms such as yeast used for fermentation.

2.3.1.6 Simultaneous saccharification and co-fermentation (SSCF)

This involves carrying out the hydrolysis and saccharification in the same unit with co-fermentation of pentose sugars. Usually, genetically modified *Saccharomyces cerevisiae* strains that can ferment xylose are used since normal *Saccharomyces cerevisiae* cannot ferment pentose sugar [47]. Like SSF, SSCF has the advantages of lower cost, higher ethanol yield and shorter processing time [43]. In addition, SSCF helps to minimize the inhibition caused by sugars during the enzymatic hydrolytic process and increases xylose to glucose concentration ratio as most of the microorganisms consume xylose.

2.3.1.7 Consolidated bioprocessing (CBP)

This requires the enzyme production, hydrolysis and fermentation to be carried out in a single unit. The microorganism mostly used in this process is *Clostridium thermocellum* as it has the capacity to synthesize cellulase which degrades lignocellulose to monomeric sugars and produce ethanol [48]. Although, CBP is still at its nascent stage, the following advantages have been identified: less energy intensive, cheaper cost of enzyme, low cost of investment, less possibility of contamination.

2.3.2 Factors affecting bioethanol production

The factors which impact the bioethanol production include temperature, sugar concentration, pH, fermentation time, agitation rate and inoculum size [49]. High temperature could denature the enzymes and reduce their activity. The ideal temperature for the fermentation of biomass is 20–35°C [50]. The optimum yield of bioethanol production could be achieved using a concentration of 150 g/L [49]. The pH of the broth also affects the production of bioethanol because, it impacts on the bacterial contamination, yeast growth, fermentation rate and by-product formation. The optimum range of pH for the fermentation of the biomass using *Saccharomyces cerevisiae* is 4.0–5.0. When the pH is less than 4.0, a longer incubation period is required and at a pH above 5.0, ethanol concentration is significantly reduced. To optimize the yield of bioethanol, another factor to be considered is the agitation rate. The higher the agitation rate, the higher the quantity of ethanol produced. For fermentation using yeast cells, the commonly used agitation rate is 150–200 rpm. Excess agitation rate may limit the metabolic activities of the cells [49].

2.3.3 Integrated processes (IP)

This involves combining one or more processes in the bioethanol production processes from the lignocellulosic biomass for the purpose of optimization, resulting to the increase in yield and minimum production cost [33]. An example of IP is membrane reactor where both reaction and separation of products occur simultaneously [33]. The hydrolysis and fermentation processes can be integrated into separate hydrolysis and fermentation (SHF) and simultaneous saccharification and fermentation (SSF). As discussed earlier, SHF provides an opportunity for the temperatures of the cellulases and *Saccharomyces* cerevisiae to be controlled separately for the efficient operation of each process. Operating the SHF at the optimum temperatures of 45–55°C for cellulase and less 32°C *Saccharomyces* cerevisiae provides favourable conditions for the pentose and hexose sugars to be fermented in a single-step process giving rise to a method known as separate hydrolysis and co-fermentation (SHCF), see **Figure 7** [52]. As mentioned earlier, one of the disadvantages of the enzymatic hydrolysis is the inhibition of the cellulase caused by high concentration of glucose produced. This challenge can be solved by increasing the concentration of the enzyme or by using SSF [52]. SSF allows glucose obtained from the enzymatic hydrolysis to be converted directly to ethanol through fermentation in the same reactor. Some investigators have argued that SSF process is rather sequential and not simultaneous. Thus, saccharification coupled with co-fermentation (SCCF) is used (see **Figure 8**) [50].

The SSF has been developed further to a technology known as consolidated bioprocessing (CBP) by integrating enzyme production into the operation (**Figure 9**). As mentioned earlier on, enzyme production, hydrolysis and fermentation are conducted in a single unit [48].

2.4 Ethanol recovery

The fermentation of monomeric sugars is usually followed by ethanol recovery from the fermentation broth. Usually, the water content of the broth is reduced to approximately 0.5% by volume enabling the formation of anhydrous ethanol with a minimum of 99.5% by volume. This operation is constrained by the azeotropic nature of ethanol-water solution and can be carried out based on the principle of distillation (i.e. leveraging the difference in boiling point of the components of the solution). The problem with the azeotropic solution is overcome by using a separating agent which alters the relative volatility of the key component. The techniques used in the recovery of pure ethanol from the fermentation broth include adsorption distillation, azeotropic distillation, diffusion distillation, extractive distillation, vacuum distillation, membrane distillation and chemical dehydration. The conventional techniques include azeotropic distillation, liquid-liquid extraction and extractive distillation [53]. Extractive distillation is the most predominantly used for large scale operations. There are some other techniques that are gaining traction for future use especially due to less energy requirement. These are pervaporation and salt distillation [54].

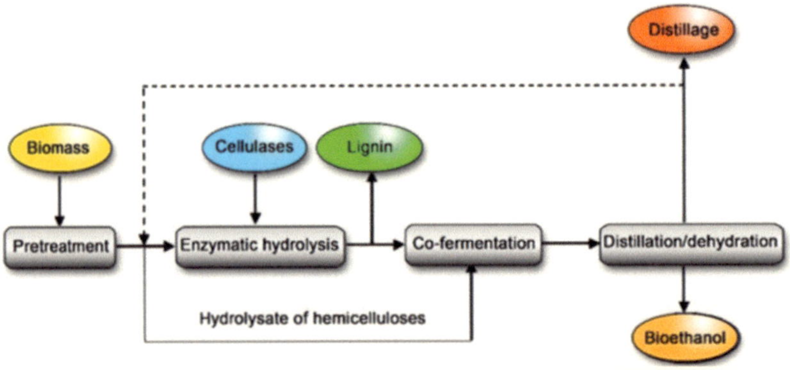

Figure 7.
Separate hydrolysis and co-fermentation (SHCF) [51].

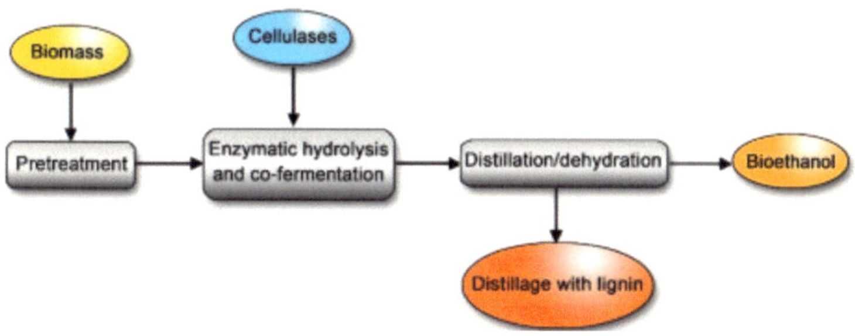

Figure 8.
Saccharification coupled with co-fermentation (SCCF) [51].

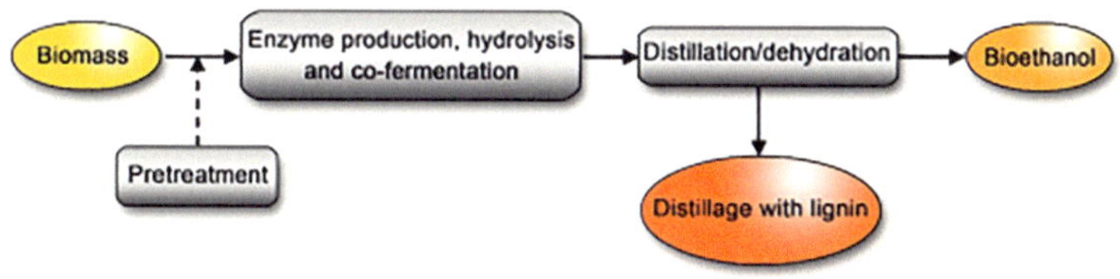

Figure 9.
Consolidated bioprocessing [51].

3. Life cycle assessment (LCA)

This assessment is usually carried out to measure the environmental impact of bioethanol production using different feedstocks. The LCA tool helps for the identification of potential impacts during a process design and for decision making in order to improve the process prior to scaling-up [55]. The LCA methodology consists of four main stages including definition of goal and scope, Life Cycle Inventory Analysis (LCIA), Impact assessment, and Interpretation of the results [1]. The LCIA can be conducted using methodologies such as CML 2002, Eco-indicator 99, ReCiPe, LIME, Lucas and TRACI depending on the impact categories and selection of indicators [56]. Numerous investigations have been conducted on the environmental impact of bioethanol and allied chemical products from different lignocellulosic feedstock (**Table 3**). The table indicates that bioethanol has the capacity to reduce the greenhouse gases emission and global warming potential substantially and hence facilitates the protection of the ozone layer.

4. Techno-economic analysis (TEA)

TEA is an effective tool used in assessing the economic feasibility of different processes pertaining to bioethanol production. This analysis provides the opportunity to evaluate the technical and economic efficiencies of different process routes leading to bioethanol production with an overarching objective of choosing the best route(s) [60]. The technical aspect of the analysis involves the development of the process flow diagram, and rigorous material and energy balance calculations using simulation software such as Aspen Plus and SuperPro. The economic aspect involves the capital and project cost estimation, discounted cash flow and determination of the minimum ethanol selling price (MESP). This may be carried out using the Aspen Economic Evaluator package (Aspen Technology, Inc., USA). The MESP can be used for comparing the differences in technology between processes

Feedstock	Method of production	Environmental analysis		Ref.
		Method of Assessment	Main impacts	
Cattle manure (CM)	Drying, milling, pretreatment, solid phase separation, Separated Hydrolysis and Fermentation (SHF), and distillation	SimaPro software v.7.3.2 was used with ReCiPe method and EcoInvent libraries	1. Results from midpoint indicators with normalized data showed that the main impacts were on human toxicity, freshwater eutrophication, terrestrial and marine ecotoxicity and fossil depletion 2. Endpoint indicators showed that the main impacts were climate change, human toxicity, particulate matter formation and fossil resource depletion	[55]
Wheat straw	Steam pretreatment, hydrolysis, Fermentation, distillation, enzyme recycling, C_5 sugars drying, and lignin pelletizing	Simplified LCA approach according to European Renewable Energy Directive (RED)	Up to 87% GHG potential mitigation	[57]
Sweet potato	Cultivation, and conversion of sweet potato to bioethanol	EcoInvent 3.1database, literature and field data. SimaPro software was used for the impact assessment with CML IA baseline 3.02 method	Reduction of global warming potential (GWP) of 44%	[58]
Loblolly pine, eucalyptus, unmanaged hardwoods, forest residues, and switchgrass	Thermochemical conversion	SimaPro 7.3 was used with the US Life Cycle Inventory dataset. The Tool for the Reduction and Assessment of Chemical and Other Environmental Impacts (TRACI) impact assessment method was used to calculate the life cycle environmental and human health midpoint impacts	Reduction in the GHG emissions by more than 60% compared to gasoline.	[59]

Table 3.
LCA of bioethanol production from different feedstock.

or for carrying out sensitivity analyses which helps to determine where economic or process performance improvement is required. Numerous investigators have studied the techno-economic analysis of bioethanol production using lignocellulose as feedstocks. For instance, Quintero et al. conducted a techno-economic analysis of bioethanol production from sugar cane bagasse, coffee cut-stems, rice husk, and empty fruit bunches for the Colombian case [61]. These researchers used Aspen Plus and Aspen Process Economic Analyzer for the process simulation and economic analysis, respectively. The results obtained showed that considering the four lignocellulosic biomasses assessed, the production cost of bioethanol from the empty fruit bunches was the lowest (0.49 US$/L).

5. Exergy analysis

Exergy is the maximum amount of work that can be obtained when a mass or energy stream is brought to equilibrium with a reference environment. Exergy analysis helps in: identifying the location, source, and the magnitude of true thermodynamic losses; determining the exergy losses in each process step which reduces the performance of the system, and comparing various process configurations to determine the most efficient route for maximum productions. Exergy analysis can be used to evaluate the thermodynamic efficiency (η) of a system, Eq. (2) [62].

$$\eta = \frac{Exergy\ of\ useful\ products}{Input\ Exergy} \tag{3}$$

Eq. (2) can be adapted to evaluate the overall efficiency (η) of the production of ethanol through biochemical process, Eq. (3)

$$\eta = \frac{\mathrm{E_{X,et} + P_{net} + E_{X,res}}}{\mathrm{E_{X,bm} + \Sigma E_{X,ch} + E_{X,LT}}} \tag{4}$$

Where; $E_{X,et}$ = chemical exergy of ethanol, P_{net} = net electricity produced by the system, $E_{X,res}$ = exergy of the lignin-enriched residue, $E_{X,bm}$ = input chemical exergy of biomass, $\Sigma E_{X,ch}$ = sum of the chemical exergies of all inputs to the process, and $E_{X,LT}$ = exergy of a potential low temperature heat source supplied to the system

Exergy balance can be applied to the system boundary of a unit operation of a process to evaluate the thermodynamics losses, Eq. (4). This equation shows that contrary to energy, exergy is not conserved.

$$\sum_{\mathrm{in}} \mathrm{E_X} = \sum_{\mathrm{out}} \mathrm{E_{X,prd}} + \sum_{\mathrm{out}} \mathrm{E_{X,wstprd}} + \mathrm{I} \tag{5}$$

Where; $\sum_{\mathrm{in}} \mathrm{E_X}$ = total input exergy flow; $\sum_{\mathrm{out}} \mathrm{E_{X,prd}}$ = total output exergy flow in the products; $\sum_{\mathrm{out}} \mathrm{E_{X,wstprd}}$ = total output exergy flow in the waste products from the unit processes, and I = exergy destruction due to internal irreversibility ($\mathrm{I} \neq 0$ for an irreversible process).

From Eq. (4) exergy loss associated with the unit process = $\sum_{\mathrm{out}} \mathrm{E_{X,wstprd}} + \mathrm{I}$.

The exergy analysis can be combined with life cycle assessment (LCA) to form exergetic life cycle assessment (ELCA) which helps to account for all environmental issues as well as the depleting natural resources [62]. This involves closed material and energy balances and can be carried out by determining the exergy destruction during the process.

Several works have been carried out by investigators in the area of applying the exergy tools to evaluate the process performance in the bioethanol production from biomass. Hurtado et al. used exergy analysis to evaluate the efficiency of the bioethanol production processes using rice husks as feedstock [63]. Aspen Plus software was used to simulate the process and the results of the exergy analysis showed that the pretreatment stage required improvement of either mass or energy as the stage gave the lowest exergetic efficiency and highest irreversibilities.

6. Process simulation

This is the pictorial representation of chemical, physical, biological, other technical processes and unit operations in a simulation software. The software helps: in the design of environmental-friendly and safe processes, reduction of capital and operating costs, to provide functionality and flexibility needed for modelling efficient biofuel processes, to enhance heat recovery processes, reconcile data, verify operating conditions, efficient and optimal process design, regulatory compliance, and operational analysis of the biofuels process [64]. With the simulation software, engineers can work virtually, thereby avoiding expenses and time delays associated with testing the process in the real world [64]. Examples of simulation software used in simulating the bioethanol production include Aspen plus, Chemcad, Prosimplus, Hysys and PRO/II [1, 64]. The most commonly used software in biorefinery is the Aspen plus [1]. The simulation of bioethanol production from lignocellulosic biomass requires interconnecting the various unit operations: pretreatment, hydrolysis, fermentation and distillation involved. This has been demonstrated by Peralta-Ruíz et al. by simulating the bioethanol production process using residual microalgae biomass as the feedstock [65]. These investigators evaluated the most effective route from three technologies: simultaneous saccharification and co-fermentation (SSCF), simultaneous saccharification and fermentation (SSF) and separate saccharification and fermentation using acid hydrolysis (SHF) leading to the highest yield of bioethanol. The simulation was carried out using Aspen Plus 7.1 and the results obtained showed that SSCF gave the highest yield of 23.6% and SHF the lowest yield of 18.5%. With these results, they concluded that enzymatic technologies could be used for microalgal production of bioethanol.

7. Conclusions

The quantity of bioethanol produced globally is increasing (110 billion liters in 2018 and could be 140 billion liters in 2022) with US and Brazil currently the highest producers. These countries produce bioethanol (1G bioethanol) from corn and sugar cane, respectively. Due to high cost of production with 40–70% contribution from the feedstocks, other sources of feedstocks are being considered leading to the production of the 2G, 3G and 4G bioethanol, respectively. The high cost of processing lignocellulosic biomass into bioethanol still makes the route unattractive compared to 1G bioethanol. The 2G bioethanol constitutes less than 3% of the total bioethanol production and has a higher GHG reduction potential compared to 1G bioethanol. Results show that bioethanol has the capacity to reduce the greenhouse gases emission and global warming potential substantially and hence facilitates the protection of the ozone layer.

In order to increase the yield of bioethanol and minimize the cost of production, different processes maybe combined through integrated processes, for example SHCF, SCCF and CBP.

Based on the LCA of the bioethanol production, the environmental impacts depend on the feedstock availability and the technology used for converting them to bioethanol.

Author details

Ifeanyichukwu Edeh
Chemical Engineering Department, Faculty of Engineering, University of Port Harcourt, Nigeria

*Address all correspondence to: ifeanyichukwu.edeh@uniport.edu.ng

References

[1] Sharma B, Larroche C. Comprehensive assessment of 2G bioethanol production. Bioresource Technology. 2020; 1-9.

[2] Mohanty SK, Swain MR. Bioethanol production from corn and wheat. In: Ramesh CR, Ramachandran S (Eds.). Bioethanol Production from Food Crops. Academic Press; 2019. pp. 45-59.

[3] Ethanol Basics (Fact Sheet). Clean cities, energy efficiency & renewable energy (EERE). https://www.afdc.energy.gov/uploads/publication/ethanol_basics.pdf; 2015.

[4] Carrillo-Nieves D, Alanís, MJR, de la Cruz Quiroz R, Ruiz HA, Iqbal HM, Parra-Saldívar R. Current status and future trends of bioethanol production from agro-industrial wastes in Mexico. Renew. Sustain. Energy Rev. 2019;102: 63-74.

[5] Dahman Y, Syed K, Begum S, Roy P, Mohtasebi B. Biofuels: Their characteristics and analysis. In: Verma D, Fortunati E, Jain S, Zhang X, editors. Biomass, Biopolymer-Based Materials, and Bioenergy. Woodhead Publishing, Cambridge. 2019; pp.277-325

[6] Sindhu R, Binod P, Pandey A, Ankaram S, Duan Y, Awasthi MK. Biofuel Production from Biomass: Toward Sustainable Development. In: Kumar S, Kumar R, Pandey A (Editors). Current Developments in Biotechnology and Bioengineering: Waste Treatment Processes for Energy Generation Elsevier. 2019; p. 79-92

[7] Toor M, Kumar SS, Malyan SK, Bishnoi NR, Mathimani T, Rajendran K, Pugazhendhi A. An Overview on bioethanol production from lignocellulosic feedstocks. Chemosphere. 2020; 242:1-12

[8] Anyanwu RC, Rodriguez C, Durrant A, Olabi AG. Micro-Macroalgae Properties and Application. In: Hashmi S (Ed). Reference Module in Materials Science and Materials Engineering. Elsevier B. V. 2018. https://doi.org/10.1016/B978-0-12-803581-8.09259-6

[9] Ho SH, Huang SW, Chen CY, et al. Bioethanol production using carbohydrate-rich microalgae biomass as feedstock. Bioresour. Technol. 2013.135; 191-198.

[10] Chng LM, Lee KT, Chan DJC. Synergistic effect of pretreatment and fermentation process on carbohydrate-rich Scenedesmus dimorphus for bioethanol production. Energy Convers. Manag. 2017;141: 410-419.

[11] Kim HM, Oh CH, Bae HJ. Comparison of red microalgae (Porphyridium cruentum) culture conditions for bioethanol production. Bioresour. Technol. 2017; 233:44-50.

[12] Ashokkumar V, Salim MR, Salam Z, et al. Production of liquid biofuels (biodiesel and bioethanol) from brown marine macroalgae Padina tetrastromatica. Energy Convers. Manag. 2017; 135:351-361.

[13] Jambo SA, Abdulla R, Marbawi H, Gansau JA. Response surface optimization of bioethanol production from third generation feedstock-Eucheuma cottonii. Renew. Energy. 2019; 132:1-10

[14] Maurya DPA. Singla and S. Negi. An overview of key pretreatment processes for biological conversion of lignocellulosic biomass to bioethanol. Biotechnology. 2015;3: 1-13.

[15] Bensah EC, Mensah M. Chemical pretreatment methods for the

production of cellulosic ethanol: Technologies and innovations. Int J Chem Eng. 2013;719607

[16] Ceballos RM. Bioethanol and Natural Resources: Substrates, Chemistry and Engineered Systems. Taylor & Francis Group, USA, 2018. pp. 201

[17] Brodeur GE, Yau K, Badal J, Collier KB, Ramachandran, Ramakrishnan S. Chemical and physicochemical pretreatment of lignocellulosic biomass: A review. Enzyme Res 2011; 787532

[18] Egüés IC, Sanchez I, Mondragon, Labidi J. (2012). Effect of alkaline and autohydrolysis processes on the purity of obtained hemicelluloses from corn stalks. Bioresour Technol 103(1): 239-248S.S.

[19] Hassan SS, Williams GA, Jaiswal AK. Emerging Technologies for the Pretreatment of Lignocellulosic Biomass. Bioresource Technology. 2018

[20] Zhang YHP, Ding SY, Mielenz JR, Cui JB, Elander RT, Laser M, Himmel ME, McMillan JR, Lynd. LR. Fractionating recalcitrant lignocellulose at modest reaction conditions. Biotechnol Bioeng. 2007;97: 214-223.

[21] Sathitsuksanoh N, George A, Zhang YHP. New lignocellulose pretreatments using cellulose solvents: A review. J Chem Technol Biotechnol. 2013; 88: 169-180.

[22] Sathitsuksanoh N, Zhu Z, Wi S, Zhang YH. Cellulose solvent-based biomass pretreatment breaks highly ordered hydrogen bonds in cellulose fibers of switchgrass. Biotechnol Bioeng. 2011;108: 521-529.

[23] da Costa Lopes AM, João KG, Morais ARC, Bogel-Łukasik E, BogelŁukasik R. Ionic liquids as a tool for lignocellulosic biomass fractionation. Sustain Chem Process. 2013; 1: 1-31.

[24] Alvira P, Tomas-Pejo E, M. Ballesteros M, Negro M. Pretreatment technologies for an efficient bioethanol production process based on enzymatic hydrolysis: A review. Bioresour Technol. 2010;101: 4851-4861.

[25] Joshi B, Raj M, Dinita B, et al. Lignocellulosic ethanol production: Current practices and recent developments. Biotechnol. Mol. Biol. Rev. 20111; 6:172-182.

[26] Azhar SHM, Abdulla R, Jambo SA, Marbawi H, Gansau JA, Faik AAM, Rodrigues KF. Yeasts in sustainable bioethanol production: A review. Biochemistry and Biophysics Report 10. 2017;52-61

[27] Imran M, Anwar Z, Irshad M, Asad MJ, Ashfaq H. Cellulase production from species of fungi and bacteria from agricultural wastes and its utilization in industry: a review. Adv. Enzyme Res. 2016.https://doi.org/10.4236/aer.2016.42005.

[28] MetaCyc. MetaCyc Reaction: 3.2.1.4. 2014.

[29] MetaCyc. MetaCyc Reaction: 3.2.1.91. 2014

[30] Brenda. Reaction catalyzed by beta-glucosidase (3.2.1.21), Sulfolobus solfataricus beta-glycosidase (3.2.1. B26), Pyrococcus furiosus beta-glycosidase (3.2.1. B28), glucan 1,4-beta-glucosidase (3.2.1.74), lactase (3.2.1.108). 2015.

[31] MetaCyc. MetaCyc Reaction: 3.2.1.8. 2014.

[32] MetaCyc. MetaCyc Reaction: 3.2.1.37. 2014

[33] Senatore A, Dalena F, Basile A. Novel bioethanol production processes and purification technology using membranes. In: Basile A, Centi G, Falco MD, Laquaniello G, editors. Catalysis, Green Chemistry and Sustainable Energy. 2019; 179:359-384

[34] Surendhiran D, Sirajunnisa AR. Role of Genetic Engineering in Bioethanol Production from Algae. Bioethanol Production from Food Crops. Academic Press. 2019. pp. 361-381.

[35] Ji H, Yu J, Zhang X, Tan T. Characteristics of an immobilized yeast cell system using very high gravity for the fermentation of ethanol, Applied Biochemistry and Biotechnology. 2012;168: 21-28

[36] Zeng Y, Zhao S, Yang S, Ding SY. Lignin plays a negative role in the biochemical process for producing lignocellulosic biofuels, Current Opinion in Biotechnology. 2014; 27:38-45

[37] Letti LAJ, Karp SG, Woiciechowski AL, Soccol CR. Ethanol production from soybean molasses by Zymomonas mobilis. Biomass and Bioenergy. 2012;44: 80-86.

[38] Ma R, Castro-Dominguez B, Mardilovich IP, Dixon AG, Ma YH. Experimental and simulation studies of the production of renewable hydrogen through ethanol steam reforming in a large-scale catalytic membrane reactor, Chemical Engineering Journal. 2016; 303:302-313

[39] Skory CD, Freer SN, Bothast RJ. Screening for ethanol-producing filamentous fungi, Biotechnology Letters. 1997;19; 203-206.

[40] Liu CG, Xiao Y, Xia XX, Zhao XQ, Peng L, Srinophakun P, Bai FW. Cellulosic ethanol production: Progress, Challenges and strategies for solutions. Biotechnology Advances. 2019; 37:491-504

[41] Cheng NG, Hasan M, Kumoro AC. Production of ethanol by fed-batch fermentation. Pertanika J. Sci. Technol. 2009;17:399-408

[42] Jain A, Chaurasia SP. Bioethanol production in membrane bioreactor (MBR) System: a review, Int. J. Environ. Res. Dev. 2014;4: 387-394.

[43] Chandel AK, Es C, Rudravaram R, et al. Economics and environmental impact of bioethanol production technologies: an appraisal, Biotechnol. Mol. Biol. Rev. 2007; 2:14-32

[44] Tavva SMD, Deshpande A, Durbha SR, Palakollu VAR, Goparaju AU, Yechuri VR, Bandaru VR, Muktinutalapati VSR. Bioethanol production through separate hydrolysis and fermentation of Parthenium hysterophorus biomass. Renew. Energy. 2016; 86: 1317-1323.

[45] Rastogi M, Shrivastava S. Current methodologies and advances in bioethanol production. J. Biotechnol. Biores. 2018; 1:1-8.

[46] Sakimoto K, Kanna M, Matsumura Y. Kinetic model of cellulose degradation using simultaneous saccharification and fermentation. Biomass Bioenergy. 2017; 99:116-121.

[47] Bondesson PM, Galbe M. Process design of SSCF for ethanol production from steam-pretreated, acetic-acid-impregnated wheat straw. Biotechnol. Biofuels. 2016; 9:222. https://doi.org/10.1186/s13068-016-0635-6.

[48] Hasunuma T, Kondo A. Consolidated bioprocessing and simultaneous saccharification and fermentation of lignocellulose to ethanol with thermotolerant yeast

strains. Process Biochem. 2012. https://doi.org/10.1016/j.procbio.2012.05.004.

[49] Zabed H, Faruq G, Sahu JN, et al., Bioethanol production from fermentable sugar juice, Sci. World J. 2014; 1-11

[50] Liu R, Shen F. Impacts of main factors on bioethanol fermentation from stalk juice of sweet sorghum by immobilized Saccharomyces cerevisiae (CICC 1308), Bioresour. Technol. 2008; 99:847-854.

[51] Zhao XQ, Bai FW. Zinc and yeast stress tolerance: micronutrient plays a big role. J. Biotechnol. 2012; 158:176-183.

[52] Balat M. Production of bioethanol from lignocellulosic materials via the biochemical pathway: a review, Energy Conversion and Management. 2011; 52: 858-875.

[53] Nitsche M, Gbadamosi R. Extractive and azeotropic distillation. In: Nitsche, M., Gbadamosi, R. (Eds.), Practical Column Design. Guide. Springer International Publishing, Cham. 2017; pp. 153-164. https://doi.org/10.1007/978-3-319-51688-2_5

[54] Nagy E, Mizsey P, Hancsók J, Boldyryev S, Varbanov P. Analysis of energy saving by combination of distillation and pervaporation for biofuel production. Chem. Eng. Process. Process Intensif. 2015; 98, 86-94. https://doi.org/10.1016/j.cep.2015. 10.010.

[55] Azevedo AD, Fornasier F, Szarblewski MDS, Schneider RDCS, Hoeltz M, Souza DD. Life cycle assessment of bioethanol production from cattle manure. Journal of Cleaner Production. 2017; 162: 1021-1030

[56] Rosenbaum RK, Hauschild MZ, Boulay AM, Fantke P, Laurent A, Núñez M, Vieira M. Life cycle impact assessment, in: Life Cycle Assessment: Theory and Practice. 2017; https://doi.org/10.1007/978-3-319-56475-3_10.

[57] Zech KM, Meisel K, Brosowski A, Toft LV, Müller-Langer F. Environmental and economic assessment of the Inbicon lignocellulosic ethanol technology. Appl. Energy 2016; 171: 347-356.

[58] Costa D, Jesus J, Silva V, Silveira M. Life Cycle Assessment of Bioethanol Production from Sweet Potato (Ipomoea batatas L.) in an Experimental Plant. BioEnergy Research. 2018. https://doi.org/10.1007/s12155-018-9932-1

[59] Daystar J, Reeb C, Gonzalez R, Venditti R, Kelley SS. Environmental life cycle impacts of cellulosic ethanol in the Southern U.S. produced from loblolly pine, eucalyptus, unmanaged hardwoods, forest residues, and switchgrass using a thermochemical conversion pathway. Fuel Process. Technol. 2015; 138: 164-174.

[60] Gebremariam SN, Marchetti. Techno-economic feasibility of producing biodiesel from acidic oil using sulfuric acid and calcium oxide as catalysts. Energy Conversion and Management. 2018; 171:1712-1720.

[61] Quintero JA, Moncada J, Cardona CA. Techno-economic analysis of bioethanol production from lignocellulosic residues in Colombia: A process simulation approach. Bioresource Technology. 2013; 139:300-307

[62] Kang Q, Tan T. Exergy and CO_2 Analyses as Key Tools for the Evaluation of Bio-Ethanol Production. Sustainability. 2016;8,76; doi:10.3390/su8010076.

[63] Meramo-Hurtado S, Ojeda-Degado K, Sánchez-Tuirán E. Exergy Analysis of Bioethanol Production from Rice Residues.

Contemporary Engineering Sciences. 2018; 11:467-474

[64] Mc Mullen J, Balcom L, Calder R. Simulation of the BioEthanol Process. In: Bogle IDL, Fairweather M (Eds.), Proceedings of the 22nd European Symposium on Computer Aided Process Engineering, 17-20 June 2012, London, Elsevier B.V. 2012

[65] Peralta-Ruíz, Y, Pardo Y, González-Delgad Á, Kafarov V. Simulation of bioethanol production process from residual microalgae biomass. In: Bogle IDL, Fairweather M (Eds.), Proceedings of the 22nd European Symposium on Computer Aided Process Engineering, 17-20 June 2012, London, Elsevier B.V. 2012.

Chapter 2

Ethanol Production, Current Facts, Future Scenarios, and Techno-Economic Assessment of Different Biorefinery Configurations

Jesús David Coral Medina and Antonio Irineudo Magalhaes Jr

Abstract

Currently, the continuous depletion of non-renewable resources of fuels and chemicals has promoted the research and development of different alternatives for the replacement of fossil resources as the feedstock of fuels and chemicals. At present, one of the most important biofuels in the current economy, is bioethanol, contributing to 65% of the total biofuels production. The production of bioethanol is an attractive alternative because it would be produced using indigenous and native raw material, therefore, the socioeconomic impact mainly in developing countries would be measured by the economic incomes and increase the quality of life of small and middle farmers. The first-generation ethanol production from sugarcane, corn, or beet sugar is broadly implemented at an industrial scale. However, the second-generation ethanol (2GE) is currently still in development stages, looking for different alternatives according to each region under study. The 2GE is also subject of diverse opinions about its economic viability and its real impact on the environment, especially due to the CO_2 footprint. Consequently, this chapter has presented an overview of 2GE production, the possibilities of co-production of molecules of high value-added, and their economic and environmental assessment, including CO2 release, water consumption, solid residues disposal, and economic analysis to determine the best bioethanol based biorefinery configuration.

Keywords: biorefineries, bioethanol, ethanol controversy, techno-economic assessment, environment impact

1. Introduction

At present the countries, mainly developed, are focused on energy and food security, this phenomenon has emerged in parallel with the reduction in fossil fuels. The continuous increase in the demand for fuels and food has motivated the research to new sources. The production of biofuels and bioenergy using crops or lignocellulosic material as feedstock is an emerging tendency. Bioethanol is the most critical biofuel in the current economy contributing with 65% to global biofuel production, it can play an essential role in the energy and economic security of developed and developing nations if it is produced from native biomass [1, 2].

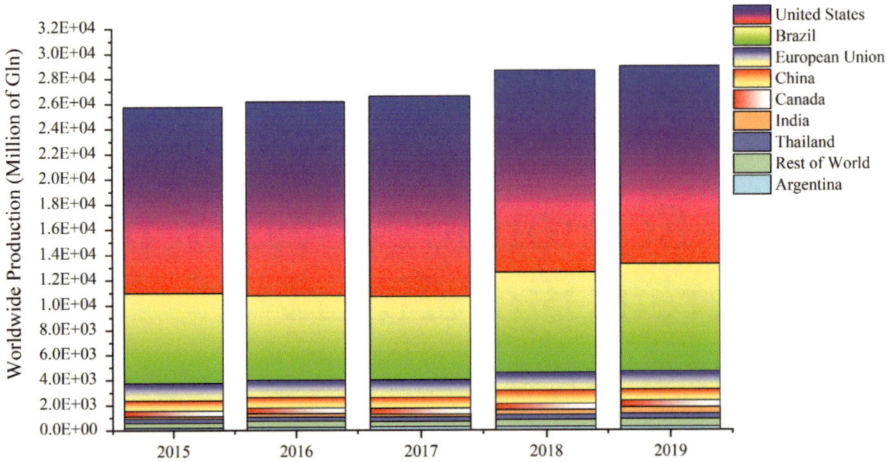

Figure 1.
Worldwide production of ethanol. Data source: https://afdc.energy.gov/data/ [3].

United States of America and Brazil leads the global ethanol production, together they produce a little more than 80% of the alcohol that is used and commercialized in the world. **Figure 1** shows the main ethanol producers in the world in millions of gallons: The United States of America, Brazil, followed by the European Union (EU), China, and Canada. In the United States ethanol is made primarily from corn, while in Brazil is produced from sugarcane. Ethanol production in the EU is exciting because even though the EU is composed of countries with high levels of technological development, the production is less than the United States and Brazil, it is probably because of the lack of standardization of feedstock. In 2014, according to the European Renewable Ethanol Report, the most widely used feedstocks to produce ethanol in Europe were corn, wheat, and sugar beet, which represent about 42, 33, and 18%, respectively [4, 5].

The production of ethanol coming from lignocellulosic material, its mean, any solid waste obtained from agro-industry, is still under study and it is subject to controversy, mainly from the technical and economic view. However, the uses of solid residues have proven to be an alternative for reducing competition for land and water available between crops for energy and food purposes [6].

This chapter has presented the status and tendencies of ethanol production using crops and lignocellulosic material, addressing environmental and economic aspects of the process, as well as future scenarios.

2. Chemistry and types of bioethanol sources

Ethanol is a relatively small chemical molecule, composed of two atoms of carbon, six hydrogens, and one oxygen, its chemical structure is C_2H_6O or C_2H_5OH to highlight the presence of the OH group. The presence of OH groups makes the ethanol a polar molecule. Moreover, the reactivity of the hydroxyl group permits its ready conversion into industrially significant products and intermediates via dehydration, dehydrogenation, condensation, etherification, and/or oxidation reactions [7].

The synthesis of ethanol can be performed both by chemical and microbiological processes. In the chemical process ethanol is produced by ethylene hydration, while the microbiological route is produced by fermentation using yeast *Saccharomyces cerevisiae* yeast, mainly [8]. In the chemical process, ethanol is manufactured by reacting to ethene with steam. The formation of the ethanol is exothermic, and the reaction is reversible. In equation (1) is presented the chemical reaction

$$CH_2 = CH_{2(g)} + H_2O_{(g)} \leftrightarrow CH_3CH_2OH_{(g)} \Delta H = -45KJ \cdot kmol^{-1} \quad (1)$$

Currently, the world ethanol production is carried out mainly by the biological pathway, referred to as alcoholic or ethanolic fermentation. During this process, sugars are converted into ethanol and CO_2 as secondary metabolites, cellular biomass, and energy. The feedstock employed is diverse, The United States of America produces ethanol from corn, Brazil bases its production process on sugar cane, the European Union from sugar beet, maize, wheat, barley, and rye. China is the fourth ethanol producer in the world, their production process is based on corn, wheat, rice, and sorghum. However, independent of the biomass, the fermentation process using hexose sugars (C_6) to produce ethanol is developed according to the equation (2).

$$C_6H_{12}O_6 + O_2 \rightarrow biomass\left(CH_{1.83}O_{0.56}N_{0.17}\right) + CO_2 + H_2O \quad (2)$$

The production of ethanol from sugar cane is one of the most important processes in South America, especially in Brazil, Colombia, and Ecuador. During this process, the sugar cane is submitted to juice extraction, with the aim to obtain a syrup rich in sugars, after which it must be sterilized to inoculate yeast, specifically *Saccharomyces cerevisiae*, which is perhaps one of the most studied and domesticated for industrial purposes.

Ethanol production using corn as feedstock requires more steps, mainly because the starch present in corn, is not metabolized directly by the yeast, therefore, is necessary to break down the starch into monomers of glucose, this is commonly carried

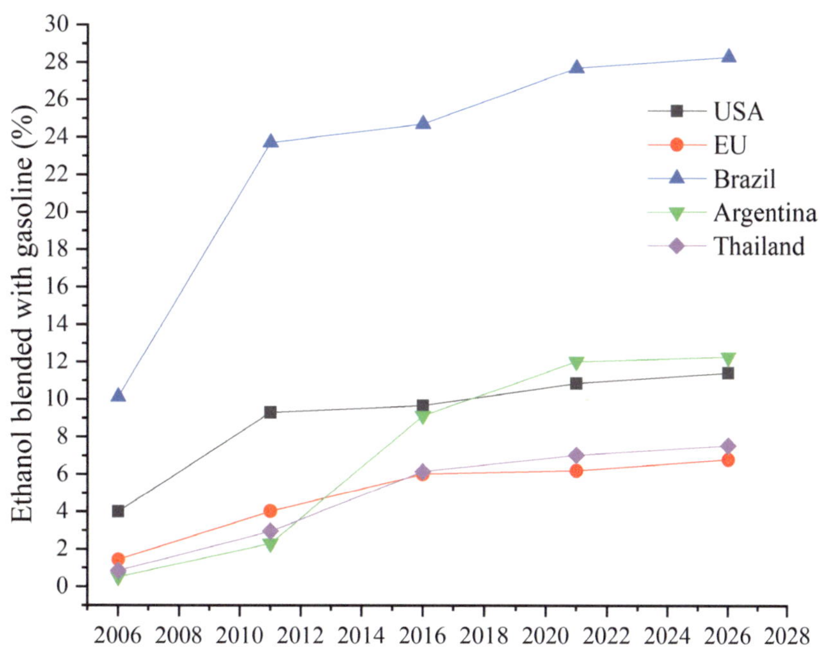

Figure 2.
Percentage of blended ethanol with gasoline in the largest producers of alcohol in the world.

out using enzymes such as alpha-amylase [9]. After monosaccharides solubilization, the fermentation process and downstream operation are like those carried out by sugarcane.

In all cases, it is mean indifferent of the feedstock used, the biomass must be pretreated with the aim to solubilize the sugar in the monomeric glucose form, for subsequent fermentation, from the fermentation broth, the water, nutrients, and salts contained in the mixture must be removed to obtain ethanol at the azeotropic point or hydrated ethanol. This operation is commonly developed by a sequence of distillation operation units.

From an economic point of view, the industry of ethanol is extremely attractive because it is used as a blending agent with gasoline and chemical building blocks. **Figure 2** shows the increased participation of ethanol in the blending with gasoline.

As is presented in **Figure 2**, the percentage of ethanol blended with gasoline has been increasing since 2006 in the main producers of alcohol in the world, except China. Brazil is the republic with the largest ratio of blending ethanol and gasoline near to 30%, regarding that in this country, the automotive industry manufactures cars with a flex engine, that is, the user can charge their vehicle with gasoline, alcohol, or a mixture of the two and the car will function normally.

Ethanol represents 90% of the total biofuels used, however, it is used as the chemical building block, different authors performed different studies about ethanol as a building block. From these studies are concluded that exist almost 12 final derivatives with high potential to be produced using ethanol as feedstock [10–12].

3. 2GE and 1GE ethanol controversy

The production of biofuels is particularly important for the reduction of the Global Warming effect and its direct consequence on climate change. However, ethanol production using different raw materials is subject to different analyses mainly by the food and fuel competition, added to different environmental, economic, and technical aspects. This section has presented an overview of the policies for ethanol production in the largest producers, their regulations, their financial aids, and production data.

First-generation ethanol (1GE) is the main liquid biofuel produced worldwide, with a global production of more than $25x10^3$ MGln. As was presented in **Figure 1**, the main producers are the United States of America and Brazil. In both countries, especially in Brazil, policies were launched and the government programs were created to promote the production and market of ethanol, mainly because of the energy crisis of the 1970s and the subsequent reduction of the dependence on imported fossil fuels.

In the United States of Amerca, 1GE is produced from corn, in this process, sugar must be produced from the starch present in corn, therefore more steps are involved during the ethanol manufacturing. To overcome this, the U.S. federal government, develop four main policies from 2002 to 2012, highlighting that alcohol production greatly expanded after the adoption of the U.S. Renewable Fuel Standard in 2005. The production increases 300 percent, passing from 4 Billion gallons in 2005 to 16 billion gallons in 2017, and is planned to reach 22 billion gallons in 2022 [13]. The policy developed by the government are listed below: (*i*) Subsidies on the feedstock used in the production of ethanol, mainly corn; (*ii*) A tax credit for blended ethanol. (*iii*) A mandate establishing a minimum volume of renewable fuel that must be blended with conventional fuels sold for transportation; (*iv*) Tariffs and other charges on imported ethanol [14].

In Brazil, two types of ethanol are used as fuel for transportation: hydrous and anhydrous alcohol. To support the ethanol industry, three main regulations were developed: (*i*) a mandatory blending of anhydrous ethanol in gasoline; (*ii*) a lower tax rate for hydrous ethanol than for gasoline; (*iii*) the Brazilian government's control over the price of gasoline another policy that has a major effect on the ethanol market. Probably the national program that includes all these policies is Proalcool Program, launched in 1974 with the aim to improve sugarcane harvesting, especially in the Sao Paulo State. One of the most important objectives of Proalcool was to guarantee fair competition and the equality of prices of alcohol with respect to sugar, paying for every 48 liters of anhydrous alcohol fuel the same value as that of a 60 kg bag of "standard" crystal sugar [15].

Then a summary of policies established in the United States and Brazil are presented in **Table 1**.

As is broadly summarized in **Table 1**, if the ethanol industry has not subsidies, the commercial price is difficult to be competitive against gasoline. This is one of the reasons because ethanol production currently generates controversy.

Moreover, 1GE production is the subject to study, mainly about the present and future competition between energy production and food consumption, which may lead to an increase in the prices of agricultural commodities, consequently causing famine in countries [16, 17]. Although this discussion is not new in the international agenda of bioenergy, this still highly controversial and generates a lot of discord. The issue of "turning food for the poor into fuel for the rich" [18].

Figure 3 shows the increase in the price of sugar and corn, in parallel with the production cost of ethanol in the United States and Brazil. Besides, according to the

Concept	U.S.	Brazil
Program Objective	**Reducing the nation's dependence on imported fossil fuels**	**Reducing the nation's dependence on imported fossil fuels**
Characteristics	1. Subsidies on feedstock used in the production of ethanol	1. A mandatory blending of anhydrous ethanol in gasoline
	2. Tax credit for blended ethanol	2. A lower tax rate for hydrous ethanol than for gasoline
	3. A law establishing a minimum percentage of Ethanol that must be blended with conventional fuels	3. Control over the price of gasoline
	4. Tariffs and other charges on imported ethanol.	
Value of the tax exemption/credit	US$0.54 per gallon (1990 to 20104)	R$0.28 per liter (2002-2007)
	US$0.51 per gallon (2005 to 2009)	R$0.18 per liter (2008)
	US$0.45 per gallon (2009 to 2011)	R$0.23 per liter (2009-2010)
		R$0.15 per liter (2011)
Charge of importation	US$0.54 per gallon of ethanol	R$0.0 for 187.5 million liters of imported ethanol
	2.5% of the import value	

Table 1.
Summary of the main policies in the two largest producers of ethanol.

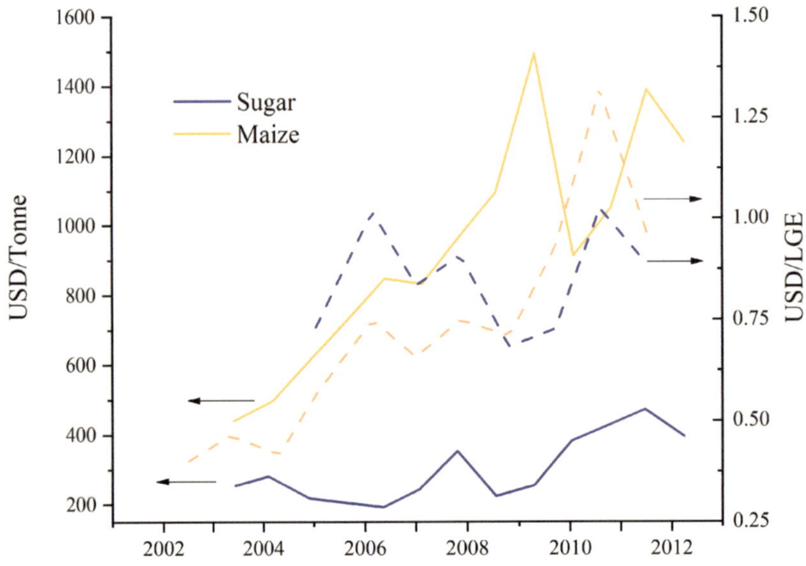

Figure 3.
Increase in the prices of corn and sugar according to the price of the gasoline equivalent (LGE).

data presented in **Figure 3**, it is well established that the price of sugar and corn has been increasing parallel to the increase in biofuels production.

To improve the environmental impact and avoid land and water competition between food and fuels, lignocellulosic material (LC) has been proposed as a viable alternative to produce liquid biofuels. LC is mainly composed of the three largest biomolecules, cellulose, hemicellulose, and lignin, their composition is a strong function of the type of biomass, harvesting time, final disposal of residues, and characterization method.

The most abundant residues are sugarcane bagasse (SCB), Oil Palm Empty Fruit Bunches (OPEFB), wheat straw, rice straw, corn straw, and soybean bagasse, are a great source of organic carbon, which would be converted to chemicals and biofuels. The LC is usually composed of cellulose, hemicellulose, and lignin. The plant matrix is organized in such a way that a microbial attack is avoided, allowing the transport of water and nutrients. This characteristic makes this type of biomass very recalcitrant, therefore, are necessary one or more steps to remove lignin, hemicelluloses to obtain fiber-rich cellulose, these steps are commonly called pretreatment of biomass, which can be chemical, thermochemical, or biological [19–21].

Currently, to improve the chain value of LC transformation is recommended the valorization of all fractions obtained during pretreatment, its mean, fractionated lignin in the form of black liquor, pentoses, mainly constituted by xylose and arabinose coming from hemicelluloses, and the glucose obtained by the enzymatic hydrolysis of cellulosic pulp. The lignin is commonly concentrated and used as an energy source to recover part of the energy expended in the process. Hemicelluloses traditionally, alternatives such as ethanol, xylitol, and furfural production were discarded. However, alternatives such as the production of ethanol, xylitol, for biological or chemical routes have been gaining attention from research centers and universities. The cellulose transformation to ethanol or some organic acids, such as citric, is probably the most extended and broadly studied and well established almost at a laboratory scale [22–24].

One of the bottlenecks of LC transformation into chemicals of high value-added and ethanol is the pretreatment, because from the economic point of view, at least 30% of the total capital investment (TCI), is required to build the pretreatment unit. From the technical view, homogeneous catalysis using acid and/or alkaline agents has proved to be the most suitable path, however, the generation of degradation products unidentified, then, commonly called humus, the formation of inhibitory compounds for biological transformation such as furfural, 5-HMF, acetic acid, mainly, and finally the low yields, almost 30% of the initial biomass is lost, makes this process a challenge for engineers and academics.

Several studies have been reported different strategies to optimize pretreatments, using steam explosion, sequential acid/alkaline, biological transformation, Ionic liquids, etc. However, in all cases, except for the biological one, biomass losses are considerable, the use of water and catalysts abundant, which makes these processes, from the environmental and economic point of view, subject to controversy [21, 25–33].

Added to the problems described below, in the case of ethanol production, two important aspects must be considered before developing a productive process. First, is the low yield obtained per kg of biomass treated, approximately by 1kg of biomass processed, are produced 100 g of ethanol would be produced. At this point is important to highlight other problem; the water present during the enzymatic saccharification and fermentation, makes the downstream process expensive and energy-intensive, therefore to develop a sustainable biorefining process, the ethanol production must be coupled with the production of high value-added molecules such as xylitol, furfural, organic acids, 5-HMF, and lignin valorization as reported by different authors [27, 34–37].

The discussion presented below had the intention of presenting the reader with an overview of the production of first and second-generation ethanol, its advantages, and associated problems, for the reader to generate their own conclusions.

4. Economic and environmental aspects of 2GE and 1GE

The 1GE production presents the two largest aspect subjects to controversy. First is the competition of land and water for crops intended for human consumption or fuel production. To overcome this problem, it has been suggested that ethanol production be carried out from lignocellulosic material from the processing of cereals, wood, oilseeds, and in general any type of biomass that is not suitable for human consumption. This section is discussed the general aspect of the economic and environmental impact of the production of both, 1GE and 2GE.

The 1GE production, especially in Brazil and the United States, has been focused on promoting rural development, with small farmers as the main beneficiaries. Based on the governmental policies and subsidies promoting ethanol production, the 1GE industry is well established and represents a market size estimated at USD 86.04 billion in 2020 (Before the sanitary emergency occasioned by SARS Covid-19). And is expected annual growth of 4.8% from 2020 to 2027 [10]. In a 1GE industry, moreover, than ethanol, exist the distilled grains, rich in carbohydrates, lipids, and protein, which can be used as byproduct to improve the economical profit [38].

On the other hand, it is 2GE, despite a promising alternative to reduce the greenhouse effect, this feedstock is not food competitive, currently is not a well-established industry with the largest production volume. Brazil is probably the country with the most advanced technology for the transformation of lignocellulosic material into bioethanol, using as feedstock the sugarcane bagasse and trash obtained from the processing of sugarcane for ethanol or refined sugar production [39].

The 2GE production using different feedstocks is broadly studied and numerous scientific reports are presented every year. From these reports, is clear that monumental efforts using various strategies have been envisioned to discover the best pretreatment method for converting biomass into fermentable sugars [40].

With the aim to develop a feasible process from the economic and energetic perspective, the two largest strategies have been reported. First is the combination of different pretreatment techniques, such as acid/alkaline pretreatment, steam explosion, CO2 explosion, ammonia fiber explosion, hydrogen peroxide treatment, Ionic liquids, ultrasonic, microwaves, and biological treatments have been studied, with two purposes mainly; first obtained the highest yields of sugar and second, the cost reduction. Parallel to these, the valorization of each fraction obtained during pretreatment, its mean, pentoses, lignin, and hexoses, for the production of molecules of high value-added, such as xylitol, furfural, 5-HMF, levulinic acid, succinic acid, fractionated lignin, to name few examples. The second strategy under study is the discovery (by isolation from nature or genetic manipulation) of robust strains that have excellent abilities to ferment hydrolyzed sugars with high yields and the largest tolerance to ethanol concentration. This strategy also seeks strains capable of fermenting pentoses for ethanol production with the objective of increasing the overall yield of alcohol production [41, 42].

From the economic evaluation of ethanol production reported in the literature [27, 43–46] is well established the following conclusions:

- The installation of a biorefinery process stand-alone using as feedstock lignocellulosic biomass is not feasible.

- The production of ethanol without valorization of the other fractions obtained during pretreatment stages is not recommended from the economic perspective.

- The Net Energy Value would be negative for the 2GE production, this means that more energy is used in the process than can be delivered through the sale of alcohol.

From the literature review and research did, we have established that the pretreatment unit is the most intensive process in economic and energetic terms, which makes the ethanol production from lignocellulosic biomass not feasible. However, the valorization of all fractions obtained during pretreatment, it would improve the economic profit and make the biorefinery process feasible from the economic point of view.

5. Future scenario for ethanol production

The development of new genetically modified strains will be one of the main advances in ethanol production. The traditional strains of *S. cerevisiae* used in the production of 1GE continue to be studied to increase yield, productivity, and tolerance to stress [47, 48]. New cultivation techniques are also being developed with the implementation of *S. cerevisiae* flocculant strains [49]. The use of these engineered strains allows the fermentation to continue because the microorganism has the capacity of auto-flocculation, settling at the bottom of the tank, and allowing higher productivity of ethanol.

The advanced strains of *S. cerevisiae* can metabolize mainly C6 sugars, such as glucose and fructose. However, from a 2GE perspective, lignocellulosic biomass can generate both C6, from the cellulosic fraction, as well as C5, as xylose and arabinose present in the hemicellulosic fraction. Thus, the use of lineages that do not have the capacity to synthesize C5 sugars with the same efficiency as C6 is a challenge to produce 2GE. Some yeast species are naturally capable of producing ethanol from xylose, such as *Candida shehatae*, *Pichia* (*Scheffersomyces*) *stipitis*, and *Pachysolen tannophilus* [50]. The use of two separate fermentation, one using cellulose hydrolysate and the other with hemicellulose hydrolysate may be the most viable alternative. Since C6 processing is already optimized with *S. cerevisiae*. Another problem with hydrolyzed broths is the presence of inhibiting compounds, such as acetic acid, furfural acid, and HMF, which can inhibit both the growth and viability of yeast and the metabolism of converting glucose to ethanol. Thus, strains resistant to inhibitory compounds are pivotal for the implementation of a biorefinery.

There is a great economic and environmental trend in the reuse of processing waste, such as lignocellulosic biomass. The productive chain of the sugar-alcohol industry can be considered the closest to a biorefinery concept. Since many wastes and by-products are no longer seen as disposable, but rather as new raw materials, impacting the price of sugar and ethanol. For example, in factories that use sugarcane, sugarcane bagasse is used to generate steam and energy by burning the residue, molasses, the by-product of the crystallization of raw sugar, in the generation of ethanol, and many investigations are carried out for the reuse of vinasse, a residue from the distillation of fermented juice, such as fertirrigation. Thus, a modern factory should contain, besides the production of sugar, the generation of bioenergy, biogas (biohydrogen and biomethane), biomolecules (organic acids, enzymes, and lipids), fertilizers, and microalgae [51].

6. Conclusions

The ethanol as biofuel is a reality, the 1GE production present a well stablished process and broadly used around the world, however, this may pose a threat to food safety. To overcome this problem, the production of ethanol using lignocellulosic material has been proposed, this appear as the most prominent alternative in terms of technological maturity. Nevertheless, the bottleneck is in the pretreatment stages, which are necessary to make fermentable sugars, therefore, standalone biorefinery process, using lignocellulosic biomass are not feasible for ethanol production. In this way, different techniques have been proposed to improve economic benefit, such as the production of value-added molecules or coupling 2GE to 1GE ethanol unit process. From the social, is well stablished that the incentives for harvesting different crops such as corn, sugarcane, wheat, rye, etc., are necessary to stimulate and benefit small producers, also, to obtain the ethanol price competitive with petroleum. Probably the environmental impact of 2GE is the most crucial, it is subject to criticism and analysis, because is not well stablished de real effect of their production, according with parameters such as CO2 liberation, water consumption, land deterioration.

Acknowledgements

The first author of this work is grateful for the financial support granted by the ceiba foundation for the postdoctoral stay in Colombia. To the Cooperative University of Colombia (Universidad Cooperativa de Colombia), campus Pasto, for being the host university.

Conflict of interest

The authors declare that there is not conflict of interest

Author details

Jesús David Coral Medina[1,2]* and Antonio Irineudo Magalhaes Jr[2]

1 Department of Industrial Engineering, Cooperative University of Colombia, Campus Pasto, ESLINGA Research Group, Nariño, Colombia

2 Department of Bioprocess Engineering and Biotechnology, Federal University of Paraná, Curitiba, Paraná, Brazil

*Address all correspondence to: jdcoralm@gmail.com

References

[1] J. Sadhukhan, E. Martinez-Hernandez, M.A. Amezcua-Allieri, J. Aburto, J.A. Honorato S, Economic and environmental impact evaluation of various biomass feedstock for bioethanol production and correlations to lignocellulosic composition, Bioresource Technology Reports. (2019) **7** 100230. https://doi.org/10.1016/j.biteb.2019.100230.

[2] M.O.S. Dias, T.L. Junqueira, O. Cavalett, L.G. Pavanello, M.P. Cunha, C.D.F. Jesus, R. Maciel Filho, A. Bonomi, Biorefineries for the production of first and second generation ethanol and electricity from sugarcane, Applied Energy. (2013) **109** 72-78. https://doi.org/10.1016/j.apenergy.2013.03.081.

[3] U.S. Department of Energy, Alternative Fuels Data Center, (2020). https://afdc.energy.gov/data/.

[4] J. Bignon, R. Wright, European Renewable Ethanol, 2015. https://doi.org/https://www.epure.org/media/1215/epure_state_industry2015_web.pdf.

[5] Sudzucker, Bioethanol as a growth market Bioethanol, 2016.

[6] A. Muscat, E.M. de Olde, I.J.M. de Boer, R. Ripoll-Bosch, The battle for biomass: A systematic review of food-feed-fuel competition, Global Food Security. (2020) **25** 100330. https://doi.org/10.1016/j.gfs.2019.100330.

[7] R.A. Dagle, A.D. Winkelman, K.K. Ramasamy, V. Lebarbier Dagle, R.S. Weber, Ethanol as a Renewable Building Block for Fuels and Chemicals, Industrial & Engineering Chemistry Research. (2020) **59** 4843-4853. https://doi.org/10.1021/acs.iecr.9b05729.

[8] M.L. Lopes, S.C. de L. Paulillo, A. Godoy, R.A. Cherubin, M.S. Lorenzi, F.H.C. Giometti, C.D. Bernardino, H.B. de Amorim Neto, H.V. de Amorim, Ethanol production in Brazil: a bridge between science and industry, Brazilian Journal of Microbiology. (2016) **47** 64-76. https://doi.org/10.1016/j.bjm.2016.10.003.

[9] Y.-C. He, Y. Ding, Y.-F. Xue, B. Yang, F. Liu, C. Wang, Z.-Z. Zhu, Q. Qing, H. Wu, C. Zhu, Z.-C. Tao, D.-P. Zhang, Enhancement of enzymatic saccharification of corn stover with sequential Fenton pretreatment and dilute NaOH extraction, Bioresource Technology. (2015) **193** 324-330. https://doi.org/10.1016/j.biortech.2015.06.088.

[10] Duffield JA, Johansson R, Meyer S, Ethanol US. An Examination of Policy. Use, Distribution, and Market Interactions: Production; 2015

[11] R.R. Noguera, S.L. Posada, Effect of the drying method on the in situ digestibility of the coffee pulp (Coffea arabica), Livestock Research for Rural Development. (2017) 1-8.

[12] J.A. Posada, A.D. Patel, A. Roes, K. Blok, A.P.C. Faaij, M.K. Patel, Potential of bioethanol as a chemical building block for biorefineries: Preliminary sustainability assessment of 12 bioethanol-based products, Bioresource Technology. (2013) **135** 490-499. https://doi.org/10.1016/j.biortech.2012.09.058.

[13] M. Kar, Second Generation Reforms, in: 2021: pp. 103-122. https://doi.org/10.4018/978-1-7998-4933-9.ch006.

[14] M. de Q.M. Jales, C.C. da Costa, Measurement of ethanol subsidies and associated economic distortions: an analysis of Brazilian and U.S. policies, Economia Aplicada. (2014) **18** 455-481. https://doi.org/10.1590/1413-8050/ea375.

[15] M.D.A. Prado Sampaio, El caso de la producción de etanol en Brasil: ¿un ejemplo para los países de América Latina?, Cuadernos de Geografía: Revista Colombiana de Geografía. (2012) **21** 147-161. https://doi.org/10.15446/rcdg.v21n1.30698.

[16] L. Tapia Carpio, F. Simone de Souza, Competition between Second-Generation Ethanol and Bioelectricity using the Residual Biomass of Sugarcane: Effects of Uncertainty on the Production Mix, Molecules. (2019) **24** 369. https://doi.org/10.3390/molecules24020369.

[17] H.I. Cobuloglu, İ.E. Büyüktahtakın, Food vs. biofuel: An optimization approach to the spatio-temporal analysis of land-use competition and environmental impacts, Applied Energy. (2015) **140** 418-434. https://doi.org/10.1016/j.apenergy.2014.11.080.

[18] M. Kuchler, B.-O. Linnér, Challenging the food vs. fuel dilemma: Genealogical analysis of the biofuel discourse pursued by international organizations, Food Policy. (2012) **37** 581-588. https://doi.org/10.1016/j.foodpol.2012.06.005.

[19] F. Wang, D. Ouyang, Z. Zhou, S.J. Page, D. Liu, X. Zhao, Lignocellulosic biomass as sustainable feedstock and materials for power generation and energy storage, Journal of Energy Chemistry. (2020). https://doi.org/10.1016/j.jechem.2020.08.060.

[20] M. Hashmi, Q. Sun, J. Tao, T.W. Jr, A.A. Shah, N. Labbé, A.J. Ragauskas, Pretreatment to Enhance Enzymatic Hydrolysis of Sugarcane Bagasse Affiliations : Department of Chemical and Biomolecular Engineering, Department of Forestry , Wildlife and Department of Microbiology, Faculty of Biological Sciences , Quaid-i-Azam Univers, Bioresource Technology. (2016). https://doi.org/10.1016/j.biortech.2016.10.089.

[21] H. Chen, J. Liu, X. Chang, D. Chen, Y. Xue, P. Liu, H. Lin, S. Han, A review on the pretreatment of lignocellulose for high-value chemicals, Fuel Processing Technology. (2017) **160** 196-206. https://doi.org/10.1016/j.fuproc.2016.12.007.

[22] T. Silva-Fernandes, L.C. Duarte, F. Carvalheiro, S. Marques, M.C. Loureiro-Dias, C. Fonseca, F. Gírio, Biorefining strategy for maximal monosaccharide recovery from three different feedstocks: Eucalyptus residues, wheat straw and olive tree pruning, Bioresource Technology. (2015) **183** 203-212. https://doi.org/10.1016/j.biortech.2015.01.136.

[23] M. Fatih Demirbas, Biorefineries for biofuel upgrading: A critical review, Applied Energy. (2009) **86** S151–S161. https://doi.org/10.1016/j.apenergy.2009.04.043.

[24] N. Smolarski, High-Value Opportunities for Lignin: Unlocking its Potential Lignin potential, (2012) 1-15.

[25] J.D.C. Medina, A. Woiciechowski, A.Z. Filho, P.S. Nigam, L.P. Ramos, C.R. Soccol, Steam explosion pretreatment of oil palm empty fruit bunches (EFB) using autocatalytic hydrolysis: A biorefinery approach, Bioresource Technology. (2016) **199** 173-180. https://doi.org/10.1016/j.biortech.2015.08.126.

[26] J.D.C. Medina, A. Woiciechowski, A. Zandona Filho, M.D. Noseda, B. Satinder Kaur, C. Ricardo Soccol, Lignin preparation from oil palm empty fruit bunches by sequential acid/alkaline treatment - A biorefinery approach, Bioresource Technology. (2015) **194** 172-178. https://doi.org/10.1016/j.biortech.2015.07.018.

[27] J.D. Coral Medina, A.L. Woiciechowski, A.Z. Filho, S.K. Brar, A.I. Magalhães Júnior, C.R. Soccol, Energetic and economic analysis of ethanol, xylitol and lignin production

using oil palm empty fruit bunches from a Brazilian factory, Journal of Cleaner Production. (2018) **195** 44-55. https://doi.org/10.1016/j.jclepro.2018.05.189.

[28] L. Tan, Y. Yu, X. Li, J. Zhao, Y. Qu, Y.M. Choo, S.K. Loh, Pretreatment of empty fruit bunch from oil palm for fuel ethanol production and proposed biorefinery process, Bioresource Technology. (2013) **135** 275-282. https://doi.org/10.1016/j.biortech.2012.10.134.

[29] R. Sun, J. Tomkinson, J. Bolton, Effects of precipitation pH on the physico-chemical properties of the lignins isolated from the black liquor of oil palm empty fruit bunch fibre pulping, Polymer Degradation and Stability. (1999) **63** 195-200.

[30] E. Palmqvist, B. Hahn-Hägerdal, Fermentation of lignocellulosic hydrolysates. I: inhibition and detoxification, Bioresource Technology. (2000) **74** 17-24. https://doi.org/10.1016/S0960-8524(99)00160-1.

[31] L.J. Jönsson, B. Alriksson, N.-O. Nilvebrant, Bioconversion of lignocellulose: inhibitors and detoxification, Biotechnology for Biofuels. (2013) **6** 16. https://doi.org/10.1186/1754-6834-6-16.

[32] M.J. Taherzadeh, K. Karimi, Pretreatment of lignocellulosic wastes to improve ethanol and biogas production: A review, 2008. https://doi.org/10.3390/ijms9091621.

[33] L. Pereira Ramos, The chemistry involved in the steam treatment of lignocellulosic materials, Quimica Nova. (2003) **26** 863-871. https://doi.org/10.1590/S0100-40422003000600015.

[34] L. Tao, A. Aden, R.T. Elander, V.R. Pallapolu, Y.Y. Lee, R.J. Garlock, V. Balan, B.E. Dale, Y. Kim, N.S. Mosier, M.R. Ladisch, M. Falls, M.T. Holtzapple, R. Sierra, J. Shi, M.A. Ebrik, T. Redmond, B. Yang, C.E. Wyman, B. Hames, S. Thomas, R.E. Warner, Process and technoeconomic analysis of leading pretreatment technologies for lignocellulosic ethanol production using switchgrass, Bioresource Technology. (2011) **102** 11105-11114. https://doi.org/10.1016/j.biortech.2011.07.051.

[35] J. Moncada, M.M. El-Halwagi, C.A. Cardona, Techno-economic analysis for a sugarcane biorefinery: Colombian case, Bioresource Technology. (2013) **135** 533-543. https://doi.org/10.1016/j.biortech.2012.08.137.

[36] T.X. Do, Y. Lim, Techno-economic comparison of three energy conversion pathways from empty fruit bunches, Renewable Energy. (2016) **90** 307-318. https://doi.org/10.1016/j.renene.2016.01.030.

[37] A.I. Magalhães, J.C. de Carvalho, J.F. Thoms, J.D.C. Medina, C.R. Soccol, Techno-economic analysis of downstream processes in itaconic acid production from fermentation broth, Journal of Cleaner Production. (2019) **206** 336-348. https://doi.org/10.1016/j.jclepro.2018.09.204.

[38] A. Chatzifragkou, D. Charalampopoulos, Distiller's dried grains with solubles (DDGS) and intermediate products as starting materials in biorefinery strategies, in: Sustain. Recover. Reutil. Cereal Process. By-Products, Elsevier, 2018: pp. 63-86. https://doi.org/10.1016/B978-0-08-102162-0.00003-4.

[39] M.O. de S. Dias, R. Maciel Filho, P.E. Mantelatto, O. Cavalett, C.E.V. Rossell, A. Bonomi, M.R.L.V. Leal, Sugarcane processing for ethanol and sugar in Brazil, Environmental Development. (2015) **15** 35-51. https://doi.org/10.1016/j.envdev.2015.03.004.

[40] S.M.R. Khattab, T. Watanabe, Bioethanol From Sugarcane Bagasse : Status and Perspectives,

in: Bioethanol Prod. from Food Crop., Elsevier Inc., 2019: pp. 187-212. https://doi.org/10.1016/B978-0-12-813766-6/00010-2.

[41] R.C. Kuhad, R. Gupta, Y.P. Khasa, A. Singh, Y.-H.P. Zhang, Bioethanol production from pentose sugars: Current status and future prospects, Renewable and Sustainable Energy Reviews. (2011) **15** 4950-4962. https://doi.org/10.1016/j.rser.2011.07.058.

[42] A. Dufey, Biofuels production, trade and sustainable Development, (2010) 153.

[43] M.D. Ferrari, M. Guigou, C. Lareo, Energy consumption evaluation of fuel bioethanol production from sweet potato, Bioresource Technology. (2013) **136** 377-384. https://doi.org/10.1016/j.biortech.2013.03.045.

[44] M.O.S. Dias, T.L. Junqueira, O. Cavalett, M.P. Cunha, C.D.F. Jesus, C.E.V. Rossell, R. Maciel Filho, A. Bonomi, Integrated versus stand-alone second generation ethanol production from sugarcane bagasse and trash, Bioresource Technology. (2012) **103** 152-161. https://doi.org/10.1016/j.biortech.2011.09.120.

[45] A. Bansal, P. Illukpitiya, F. Tegegne, S.P. Singh, Energy efficiency of ethanol production from cellulosic feedstock, Renewable and Sustainable Energy Reviews. (2016) **58** 141-146. https://doi.org/10.1016/j.rser.2015.12.122.

[46] C.N. Hamelinck, G. Van Hooijdonk, A.P. Faaij, Ethanol from lignocellulosic biomass: techno-economic performance in short-, middle- and long-term, Biomass and Bioenergy. (2005) **28** 384-410. https://doi.org/10.1016/j.biombioe.2004.09.002.

[47] S. Ostergaard, L. Olsson, J. Nielsen, Metabolic Engineering of Saccharomyces cerevisiae, Microbiology and Molecular Biology Reviews. (2000) **64** 34-50. https://doi.org/10.1128/mmbr.64.1.34-50.2000.

[48] D. Stanley, A. Bandara, S. Fraser, P.J. Chambers, G.A. Stanley, The ethanol stress response and ethanol tolerance of Saccharomyces cerevisiae, Journal of Applied Microbiology. (2010) **109** 13-24. https://doi.org/10.1111/j.1365-2672.2009.04657.x.

[49] D. Rossouw, B. Bagheri, M.E. Setati, F.F. Bauer, Co-flocculation of yeast species, a new mechanism to govern population dynamics in microbial ecosystems, PLoS ONE. (2015) **10** 1-17. https://doi.org/10.1371/journal.pone.0136249.

[50] K. Robak, M. Balcerek, Review of second generation bioethanol production from residual biomass, Food Technology and Biotechnology. (2018) **56** 174-187. https://doi.org/10.17113/ftb.56.02.18.5428.

[51] E.B. Sydney, J.C. de Carvalho, L.A.J. Letti, A.I. Magalhães, S.G. Karp, W.J. Martinez-Burgos, E. de S. Candeo, C. Rodrigues, L.P. de S. Vandenberghe, C.J.D. Neto, L.A.Z. Torres, A.B.P. Medeiros, A.L. Woiciechowski, C.R. Soccol, Current developments and challenges of green technologies for the valorization of liquid, solid, and gaseous wastes from sugarcane ethanol production, Journal of Hazardous Materials. (2021) **404**. https://doi.org/10.1016/j.jhazmat.2020.124059.

Chapter 3

Bio-Ethanol Production from Fruit and Vegetable Waste by Using *Saccharomyces cerevisiae*

Mohammad Moneruzzaman Khandaker, Umar Aliyu Abdullahi, Mahmoud Dogara Abdulrahman, Noor Afiza Badaluddin and Khamsah Suryati Mohd

Abstract

Waste from the food is a challenge to the environment all over the globe, hence there is need to be recycled. Vegetables and fruits biomass is a resource of renewable energy with significant fuel source potential for the production of electricity and steam, fuel for consumption and laboratory solvents. Bioethanol derived from biomass contributed 10–14% of the total world energy supply and solved the world crisis such as global warming and depletion of fossil fuel. Presently, bioethanol is a global issue on the efforts to reduced global pollution, contributed significantly by the petroleum or diesel combustion or combination of both. Vegetables and fruits waste significantly contains high sugar which can be utilized and serve as a raw material in the production of renewable energy using *Saccharomyces cerevisiae*. Though 80% of the current bioethanol are generated from edible materials such as starch and sugar. Biomass from lignocellulosic gathered more attention recently. The objective of this review is to account for the procedures involved in the production of bioethanol from biomass of fruits and vegetable waste through a fermentation process using *Saccharomyces cerevisiae*. In this chapter, we discussed the biomass preparation and fermentation techniques for bioethanol and reviewed the results of different fruits and vegetable waste. We found pineapple and orange fruit biomass contain a higher amount of bioethanol and easier to extract than the other fruit and vegetable wastes. Recent review coined out that dry biomass of fruit and vegetable is a promising feedstock in the utilization of bioethanol production.

Keywords: fruit, vegetable, waste, *Saccharomyces cerevisiae*, bioethanol

1. Introduction

The globe needs urgently to resort another option of sources of energy as a result of the rapid world energy supply exhaustion [1]. As a result of the depletion in oil, the world global warming and the effects of greenhouse making the earth on the condition of alarming [2]. Despite seeing the world are completely dependent on the limited sources of fossil-based petroleum that can later not withstand to meet future demands.

The world depletion fossil fuel happened, resulting in the continual price rising and the pressure for independence of oil and environments concerns lead to strong markets for biofuel [3]. The utilization of natural resources fuel leads to the vast side problem. The rapid increased of CO_2 level in the environment resulted in the global warming resulting to the negative results of the burning of fuel from petroleum-based [4]. The worlds are concern about the climatic change and the consequent need to decreasing of greenhouse emissions gasses leading to the encouragement of the usage of bioethanol as an alternative or replacement [5]. Another challenge is as a result of the arise waste dumping in an open place resulting in malignant to the natural habitat at surrounding environments of the dumpsite. The concept of producing energy in the form of a solution by utilization of the waste is affordable, cheap and efficient. Recently, an enormous number of renewable sources of energy is rapidly growing technologies of renewable energy including solid biomass, liquid fuels and biogases [6]. A biofuel is a generated fuel through biomass rather than the one produced from the formation of the geological process of oil and fossils fuel. As a result of biomass can be technically utilized directly as fuel. The term biofuel and biomass are interchangeably used. Biomass with complex or free sugar that can later form soluble sugar is used for the production of bioethanol. The feedstock is divided mostly into three major groups; starchy crops, (sugar crops and by-products of sugar refineries) and lignocellulosic biomass (LCB), they differ respectively from the sugar solutions in them [7]. Production of bioethanol from the conventional feedstock like starch-rich feedstocks (corn, potato) and sugarcane has been previously reported as the first-generation process. Nevertheless, they have economic and social barriers [8]. Bioethanol second-generation process is gaining momentum. Lignocellulosic biomass (corn stover, sugarcane bagasse, straws, stalks and switchgrass) are used for the second-generation process. One of the significant alternative processes of bio-ethanol production with easy adaptability of this biofuel to prevailing engines with better octane rating [9, 10]. Any plant material with significant amounts of sugar is utilized as a source of raw materials in bioethanol production. Sugarcane, pineapple and potato are one of the major plants that resulted in a high yield of bioethanol as byproducts due to the presence of a high amount of sugarcane in it [11] (**Figure 1**).

2. *Saccharomyces cerevisiae*

Yeast is described as basidiomycetous or ascomycetous fungi responsible for reproducing through fission or budding and formed spores which are not enclosed in the fruiting body [12]. *S. cerevisiae* is the most popular yeast in the production of ethanol due to its wide tolerance of pH making it less susceptible to infection. The ability of yeasts in catabolize six-carbon molecules is the bedrock to the production

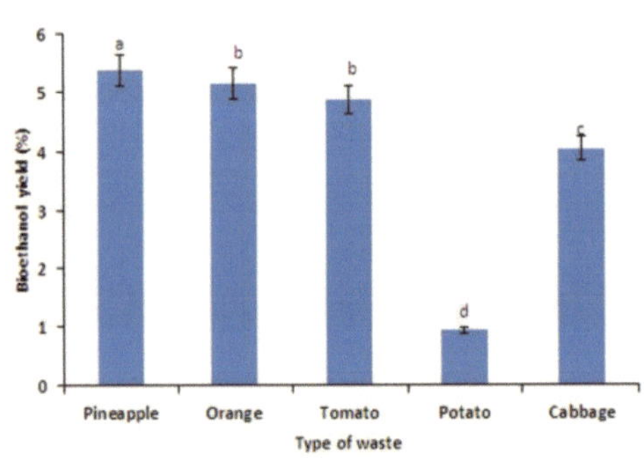

Figure 1.
The amount of bioethanol production depends on the substrate used as shown in the figure above. Adapted from Khandaker et al. [11].

of bioethanol without proceeding to the final products of oxidation which is CO_2. Diauxic shift and fermentative metabolism are the process of the production of bioethanol dependent Alcohol dehydrogenase (EC 1.1.1.1) enzymes which is encoded on the ADH1locus. During the fermentation of glucose, ADH1 catalyzes led to the production of ethanol and reduction of acetaldehyde, similarly, the reverse reaction can be catalyzed: is the process of conversion of ethanol to acetaldehyde, albeit with lower catalytic efficiency [13].

3. Fruit wastes as a source of bioethanol

Fresh citrus fruits are consumed or the citrus juice is mostly preserved which it's in ready form of consumption or concentrated form. After the extraction of citrus fruit juice, the remaining parts of the fruits serve as a rich source of lignocellulosic material and also utilized as a raw material for the fermentation of bioethanol. Simultaneous saccharification and fermentation from plantain, banana and pineapple peel through the cultured of *S. cerevisiae* and *A. niger* [14]. Different temperature (20–50°C) was used to be examined the simultaneous saccharification and fermentation of banana peels to obtain bioethanol using co-cultures of *S. cerevisiae* and *A. niger* at different pH of 4 to 7 for seven days.

The present study observed that the maximum temperature and pH for the banana peels fermentation was 30°C and 6. With these maximum conditions of temperature and pH, different concentrations 3 and 12% of yeast were utilized for performing fermentation. The study found the period for the whole fermentation to complete reduced drastically [15]. The high glucose content in pineapple and orange resulted in the excellent yield of bioethanol [11] (**Figure 2**).

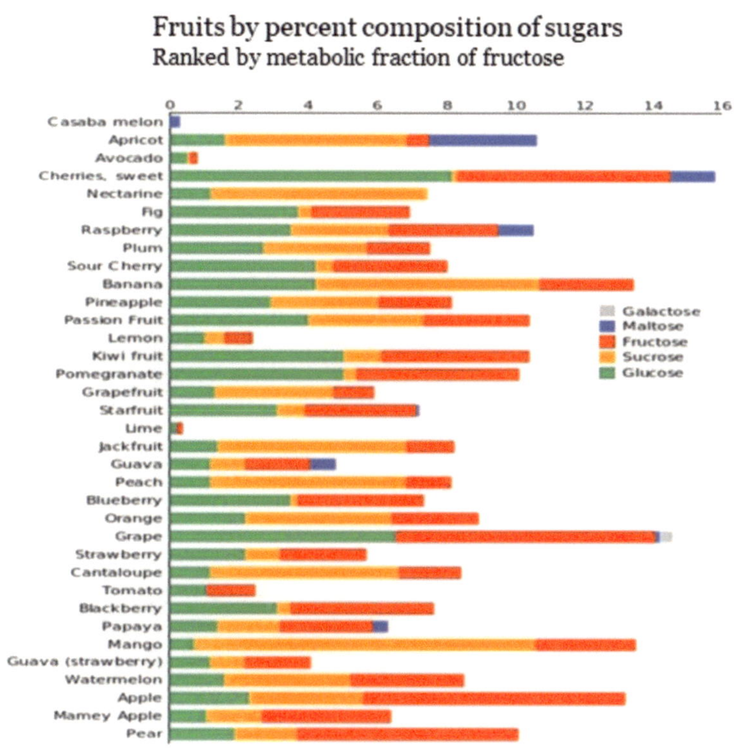

Figure 2.
Percentage of sugar composition in various fruits and vegetables [16].

4. Vegetable waste as a source of bioethanol

Rotten, peels, shells and a scraped portion of vegetables is one kind of biodegradable vegetable waste that generated in large amounts, usually dumped on ground for rotten near the household area. This act not emits an obscene odor but also creates a big irritation by attracting pigs, rats and bird as well as vectors of various human diseases. Vegetable waste mainly generates during the processing and packaging of vegetables, after preparation of cooking and post-harvest losses due to lack of storage facilities. Bioethanol can be produced through fermentation under controlled conditions. Microbial decomposition of vegetable waste generates bioethanol with high humus content. Many researchers have stated that vegetable waste is carbohydrate-rich biomass one of the potent substrates of renewable energy generations.

Research on the usage of fruit and vegetable wastes for the manufacture of biofuel is fetching attractive in different countries. Sulaiman et al. [17] abstracted a halal biorefinery for the production of bioethanol and biodiesel and value-added products in Malaysia. Vegetable wastes arise throughout the supply chain from the producer to consumer and vary widely depending on its harvesting, processing and marketing [18]. Vegetable waste can be raw, cooked, inedible and edible; parts are generated during production, harvesting, precooling, grading, storage, marketing and consumption at the consumer place. All the cut-down vegetable waste goes to landfill. Landfills spread offensive smells, produce methane which is a common greenhouse gas, and also produced a large amount of harmful leachate that can contaminate water and soil. Nevertheless, microbial digestion of vegetable waste can be used to produce bioethanol, renewable bioenergy. Vegetable waste has chemical potentials due to the high amount of saccharide in the form of lignocellulose. Promon [19] reported that vegetable waste as a high source of lignocellulose could be hydrolyzed into D-xylose and glucose.

Vegetable waste is a renowned nonedible source of lipids, amino acids, carbohydrates, and phosphates [20, 21]. All of these nonedible lignocellulose biomasses can also use for the production of bioethanol. Lignocellulose contains of 30–50% of cellulose, 20–40% of hemicellulose and lignin around 10–15% [22]. Cellulose is the main assembly of lignocellulosic built biomass which is a glucose homologous polymer associated by b-1,4 glycosidic bond [23]. After, glucose and other simple sugars production from all the sugar sources, the bioconversion endures till bioethanol is produced. Vegetable waste is widely used raw material for the production of bioethanol because it contains hemicellulose and cellulose, which can be changed into sugar by the hydrolysis method in presence of microorganisms [24]. The sugar content in vegetable waste extracts around 5% [25]. Yeast, fungi and bacteria can be used for the fermentation process [26].

5. Production of bioethanol from dry fruits and vegetable waste biomass

Pretreatment: The pretreatment is the most costly and complicated step in the conversion of LCB into ethanol. The LCB in cellulose is usually sheathed or coated by hemicelluloses resulting in hemicellulose complex cellulose that works as a chemical barrier and attacked and prevent the chances of complex enzymes under its natural condition [27]. The complexes cellulose-hemicellulose are further subjected encapsulated with signs leading to the production of physical, physical barrier to the biomass of hydrolysis to produce fermentable sugars [28].

Chemical pretreatment: Primarily acids and alkali working on the biomass of the delignification, the degree of decreasing of crystallinity of cellulose and polymerization. HNO_3, H_3PO_4, HCl and H_2SO_4 are utilized during acid pretreatment of biomass in the process the major alkali used is NaOH. Pretreatment of acid is applied in the stabilization of the fraction of hemicellulosic in the biomass, thereby making cellulose enzymes more accessible [29]. Physical pretreatments: This process convert the biomass through the increased surface accessibility area and pore volume, decreased in the degree of the polymerization of cellulose, hydrolysis of hemicellulose, partial depolymerization of lignin and its crystallinity. Physicochemical pretreatment: The exploitation of the usage of conditions and chemical compounds that affect the chemical and physical properties of the biostimulants including a large number of technologies example fiber explosion ammonia, steam exploitation, CO_2 explosion, ammonia recycling percolation wet oxidation, soaking aqueous ammonia etc. Similarly, other pretreatments methods like technologies from physicochemical also increased the accessibility area surface of the enzyme biomass, cellulose crystallinity decreased and removal of lignin and hemicellulose during pretreatment.

Biological pretreatment: Microorganisms are used are utilized particularly fungi as brown rot, white rot and soft fungi rot, the most efficient among them are white fungi rot. The above treatment became effective through the alteration of the cellulose and lignin structure and separates them from the lignocellulosic matrix. While white, soft rot and brown rot fungi attack cellulose and lignin [30].

Detoxification: Pretreatment is an important aspect of converting LCB into ethanol.

It has a significant effect on the complete process leading to the generation of lignocellulose-derived by-products under the conditions of pretreatment such as acetic acid, sugar acids, levulinic acid, formic acid, furfural and hydroxymethyl furfural acts as enzymes inhibitors for the microorganisms fermentation for the subsequent stage if the accumulation is sufficiently high [31].

Inhibitors can be checked out by:

- Chemical approach: by addition of alkali such as NaOH, reducing agents such as (sulfite, dithionite and dithiothreitol) $Ca(OH)_2$, NH_4OH, Reducing

- Treatment using enzyme: peroxidase, laccase

- Vaporization and heating: heat treatment, evaporation

- Extraction using liquid–liquid: Supercritical fluid extraction such as (Trialkylamine, supercritical CO_2), Ethyl acetate,

- Extraction using liquid–solid: Lignin, Ion exchange and Activated carbon,

- Treatments using microbes: thermospheric, *Coniochaeta ligularia*, *reibacillus* and *Trichoderma reesei* [7].

Hydrolysis: Hydrolysis is described as an industrial process where hemicellulose and cellulose present in the feedstock are converted to fermentable sugars. The fermentable sugars are maltotriose, maltose, sucrose, glucose, fructose they are generally accounting to 60–70% of the total solid dissolved. Enzymatic hydrolysis, alkaline or either acid is utilized in the conversion of cellulose and hemicellulose into their monomers sugar.

Acid hydrolysis is the oldest technology for cellulose biomass conversion to ethanol [32]. The acid hydrolysis is basically classified into two: concentrated acid hydrolysis and dilute acid. The diluted acid procedure is conducted through high pressure and temperature with a reaction time scale of one minute, reactivating continues process. The procedure of the concentrated acid utilized relatively low pressure and temperature with a much longer reaction time [33] (**Figure 3**).

Dilute acid hydrolysis the following method it is used for hydrolysis of hemicellulose and as a cellulose pretreatment to make it most accessible for the enzymes. However, both the polymers of carbohydrate are hydrolysed using acid dilution under two stages, hydrolysis process: the following stage is carrying out at a minimum temperature to utilized the hemicellulose conversion as the fraction of hemicellulose biomass for the depolymerization at a low temperature than the portion of cellulose due to the difference in the structure between these two polymers of carbohydrate [34]. The dilution of acid involved a process of a solution of sulfuric acid 1% concentration in a reactor with continues flow at a temperature of 215°C [35]. Most of the process of the acid dilution to a sugar recovery is limited to efficiency of about 50%. The most paramount challenge in the hydrolysis of acid dilution is the raising of glucose yields greater than 70% in a viable economical industrial process with a maintaining high rate of cellulose hydrolysis with minimization of decomposition of glucose. Shrinking bed reactor countercurrent technologies have been 100% success in the yielding of glucose from cellulose [36].

Concentrated Acid Hydrolysis the method provide rapid and complete cellulose of hydrolysis to glucose and sugars of hemicelluloses to 5-carbon with a little bit of degradation. The concentration of the acid process utilized mild temperature relatively, the pressure created from the pumping pressure from vessel to vessel is utilized. Dilution acid process is shorter than the reaction time [35]. Depolymerization of the cellulosic fraction is the next step. Soaking and dewatered of solid residue from the first stage was carried out in 30–40% sulfuric acid for 50 minutes. For furthering of cellulose hydrolysis is carried out at 373 k [37]. Recovery of higher sugar efficiency was the primary advantage of the concentrated acid process [38]. The process of concentrated acid offers significant cost reduction than the process of dilute sulfuric acid [39].

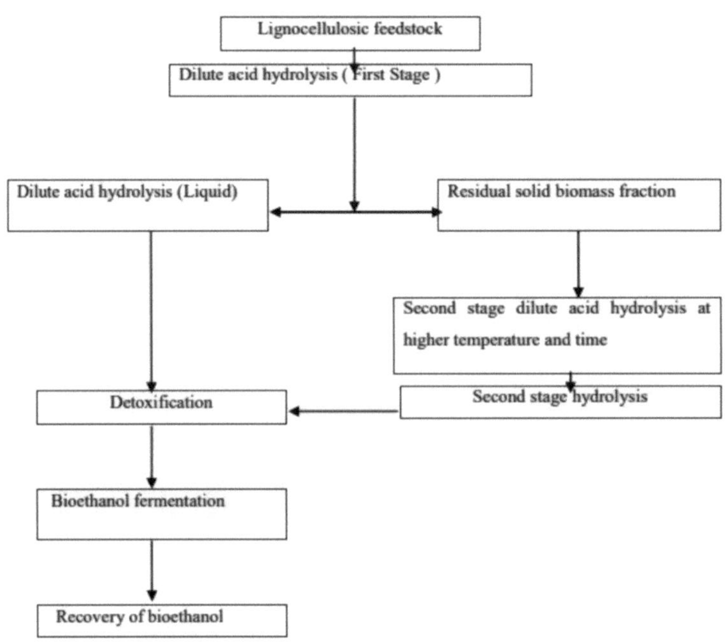

Figure 3.
Dilute acid hydrolysis flow chart of recovery bioethanol [37].

Alkaline hydrolysis the major significant from pretreatment of alkali is the removal of lignin, which greatly improved the reactivity of the remaining aspects of polysaccharides [40]. In the biomass, the aligning structure is altered by glycosidic and ester degrading side chains of the biomass through the alkaline solvents, resulting in swelling as well as cellulose decrystallization [41]. Hydrolysis of alkaline is a very slow process that requires neutralization and the recovery of the added alkali is needed. Hydrolysis of alkaline is very suitable for agricultural residue and herbaceous and woody biomass is not suitable due to its high contents of lignin [42]. Previous experiments results confirmed that hydrolysis of alkaline has the highest reaction rate, followed by hydrolysis of acid and finally degradation of hydrothermal from the glycosidic bond cleavage insoluble water carbohydrate concerned. In other to the obtained significant yield of sugar by hydrolysis of alkaline, it is very challenging as a result of dimeric and mono carbohydrates such as fructose, maltose, cellobiose or glucose are attacked severely by the temperature of alkali at 100°C [42].

Enzymatic hydrolysis for enzymatic hydrolysis to take place it required the feeds to be hydrolysed by the enzyme to become fermentable sugars. Breaking down of cellulose take place using three types of enzymes β-glucosidases, cellobiohydrolases and endo-β-1,4-glucanases. The most effective and promising among them is the enzymatic process due to the specificity of the enzyme on the substrate relatively working on the minimum temperature and generating lower inhibitors. LCB enzymatic done usually by using either microorganisms producing an enzyme that secrets directly on the enzymes during their developments in the media or enzymes system that are commercially available where the latter is widely utilized and more feasible. The commercial-scale of cost-effective ethanol its major challenge is the enzymes costs [43]. The type of biomass and the conditions of hydrolysis is the major factors dependable for the conversion of lignocellulosic biomass to fermentable sugars. Many factors are solely responsible for the yield of sugar during hydrolysis of the enzyme. The factors are generally divided into two groups. (1) factors related substrate, and interlinked with one other (2) enzymatic and factors related process. Enzymes hydrolysis is the saccharification preferred method as a result of its; high yield, high selectivity, minimum energy cost and operating milder condition than other processes [14].

Fermentation process: Bioethanol production largely depends on three processes which are simultaneous saccharification and fermentation, (SSF) and simultaneous saccharification and co-fermentation (SSCF) and separate hydrolysis and fermentation (SHF). Ethanol fermentation is completely separated lignolistic hydrolysis in SHF fermentation. Hydrolysis enzymatic separation and fermentations enabled the operation of the enzymes at a higher temperature and excellent performance. The organisms in the fermentation process operate at a lower temperature for sugar utilization optimization. SSCF and SSF fermentation and hydrolysis process occur concurrently to keep the glucose concentration low, the whole process occurs in a short process. While the SSF fermentation pentose is separated from glucose while SSCF pentose and glucose are in the same reactor [44]. Both SSCF and SSF are more efficient and preferred over the SHF as a result the operation of the later cannot be performed on the same reactor [37].

Batch, fed-batch, repeated batch or continuous mode are important technology of bioethanol fermentation. Hadiyanto et al. [45] stated that the substrate is provided at the early stages of the process without removal or addition of the medium in a batch process. The process is known as the simple system of a bioreactor with a flexible, multi-vessel and Cassy control system. In a closed-loop system with high inhibitors and sugar concentration at the beginning and ends of the fermentation is maintained and the process carried out with high product concentration [46]. Complete sterilization, require fewer labour skills, can control easily, very easy to manage feedstocks, and flexible to various product specifications are benefits of the batch system [47]. However, the productivity of the system is very low and need intensive and high labour costs. Both inhibitions of growth of the cells and production of ethanol may come from the presence of significant amount/ high concentration of sugar in the fermentation chamber [48]. However, Fed-batch fermentation overcomes the inhibition and enhanced production of ethanol. In Fed-batch fermentation, combine a form of batch and continuous modes are operated which involves increasing substrate to the fermenter devoided removing it from the medium. The size of culture in fed-batch varies significantly, but the substrate must be fed with the right component properly at a certain rate. When the low substrate concentration is maintained, higher ethanol yield in feb-batch is observed. This is because low substrate concentration permits the smooth conversion of a reasonable amount of fermentable to ethanol [47]. The benefits of this feb-batch include; higher ethanol yield, greater dissolved oxygen in the fermentation chamber, Low fermentation time and medium component exhibit a low toxic effect [48]. Fed-batch is successfully operated in non-uniform SSF system by repeatedly adding pretreated feedstock to achieve comparatively high sugar and ethanol yield [14].

Continuous operation is achieved by unceasing addition of culture medium, substrate and nutrients to bioreactor embodied active microorganisms. In continuous operation mode, the culture size is kept constant and the end products of fermentation are siphoned from the media continuously. Discrete product types such as ethanol, cells and residual sugar could be accessed from the top of bioreactor [14]. The advantages of continuous system over batch and fed-batch; small size bioreactor, higher ethanol yield and cost-effective. However, shortcomings of this technique are; the greater tendency of contamination than other types [37]. The capability of *Saccharomyces cerevisiae* to ferment and produce ethanol is drastically decreasing with longer cultivation time.

6. Characteristics/properties of bioethanol

Bioethanol fuel has the following intrinsic quality: high-octane number; this measure the engine performance (**Table 1**). The more the octane number the higher compression that the fuel can endure before ignition. Higher octane number qualifies fuel to be used in high-performance gasoline engines that need compression ratios to be high. Hence, the use of gasoline with a low octane number causes the engine knocking [49]. It drastically decreases the emission of substances that are a threat to human health eg. CO (**Table 2**). The utilization of ethanol does not employ engine modification, it does not emit CO_2, the cost of production is low, and it is eco-friendly, hence flipside of the solution to global environmental contamination [50–51].

Bioethanol fuel property	Advantages	References
High oxygen content (35% w/w)	i. Increased combustion efficiency ii. Reduced hydrocarbon and carbon monoxide emissions	[52, 53]
High octane number (107) and high latent heat of vaporization (0.91 MJ/kg)	i. Prevents premature ignition and cylinder knocking ii. Spontaneous ignition in internal combustion engines when bioethanol petrol blends are used	[54, 55]
Low energy content (21.2 MJ/dm^3)	i. Increased compression ratio ii. Decreased burn time iii. Increased power	[56, 57]

Table 1.
Advantage of bioethanol.

Bioethanol	Fossil ethanol
Renewable	Non-renewable
Waste plant material used as feedstock	Fossils source
Cost-effective	expensive
Least pollutants are released	Many pollutants are released

Table 2.
Difference between bioethanol and fossil ethanol.

7. Factors affecting bioethanol production

Temperature: the roles of temperature for *S. cerevisiae* to ferment sugar and the production of ethanol were studied. Results from previous studies show *S. cerevisiae* cells increased exponentially as the incubation begins and then get into stationary phase after prolong incubation for all operating temperatures. Experiments prove that as the temperature is progressively increasing, the time required for fermentation decreases. Nevertheless, at much high-temperature *S. cerevisiae* cells growth is inhibited and decline in ethanol production is drastic [58] (**Figure 4**). This may be due to that temperature affects the transport system or the level soluble substances and solvent in the *S. cerevisiae* cells are saturated which in turn causes the build-up of toxins ethanol inclusive inside cells [58–60].

Whereas low temperature slows the growth rate of cells which may be due to their low tolerance to ethanol at lower temperatures [62, 63].

Effect of Feedstock Concentration: feedstock encloses nutrients for microorganism's growth during the fermentation process. At high feedstock concentration, the rate hydrolysis is speed up because more compound is bound to enzymes' active site. With fixed number of enzymes and low amount of substrate cause decrease in production of ethanol because bound to enzymes' active site. A small amount of ethanol will be obtained because of low substrates bound to the enzyme's active site. Hence, the increase in feedstock concentration favors the production of ethanol [64] (**Figure 5**). However, according to Lin et al. [58] prolong exposure to a higher concentration of feedstock lead to diminishing the production of bioethanol.

Effect of pH: Fermentation process is pH sensitive. In an acidic medium with moderate pH, high ethanol production was observed (**Figure 6**). Moderately acidic

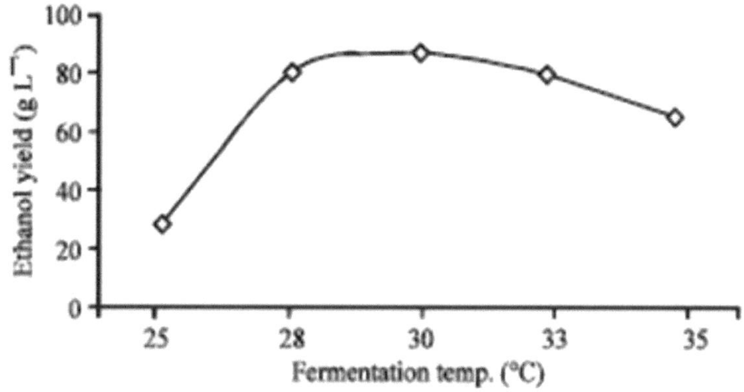

Figure 4.
Effect of temperature on bioethanol yield [61].

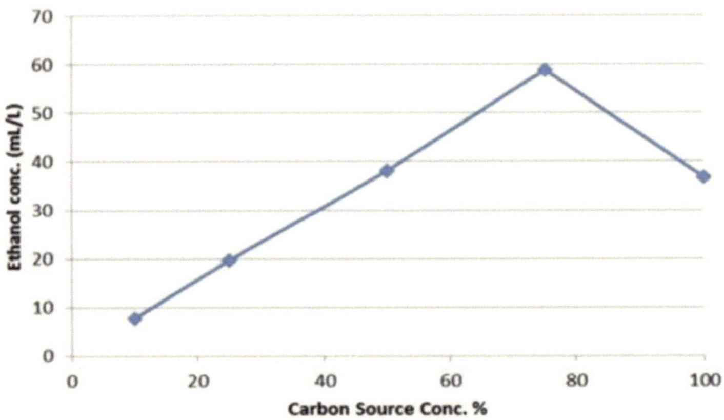

Figure 5.
Effect of feedstock on bioethanol production [65].

pH, cell permeability to some essential nutrients is influence by the concentration of H+ in the fermentation broth [28]. It has been experimentally observed that both growth and survival rate of *S. cerevisiae* is persuaded by pH in the 2.75–4.25 range. However, during fermentation for ethanol production, 4.0–4.25 is the optimum range of pH. When pH is ≤4.0, incubation period longer than necessary is required even though it does not cause a significant decrease in ethanol production. A substantial reduction of ethanol production was observed at pH above 5.0 [66, 67] (**Figure 6**).

Time of Fermentation: the rate at which growth of microorganisms occurs is affected by fermentation time (**Figure 7**). The shorter the fermentation times the more inefficient fermentation due to inadequate microorganisms growth. Equally, longer fermentation time cause affects *S. cerevisiae* growth due to high concentration of ethanol in the broth. However, using a low temperature and long fermentation result in lowest ethanol yield [28].

Agitation rate this controls to regulate the entry of nutrients from the fermentation broth to inside cells and eviction of ethanol from the cells to the fermentation broth. Higher rate of agitation leads to higher production of ethanol. It plays a role in triggering sugar takes up and the inhibition of ethanol to the cell is reduced. The frequently used agitation rate for fermentation by yeast cells is 150–200 rpm. It is inadvisable to use excess agitation rate as it reduces metabolic activities of the cell and hence, unsuitable for smooth production of ethanol [28].

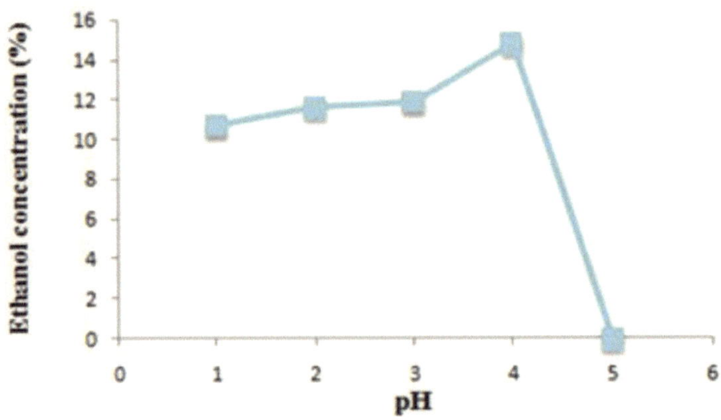

Figure 6.
Effect of pH on bioethanol production [57].

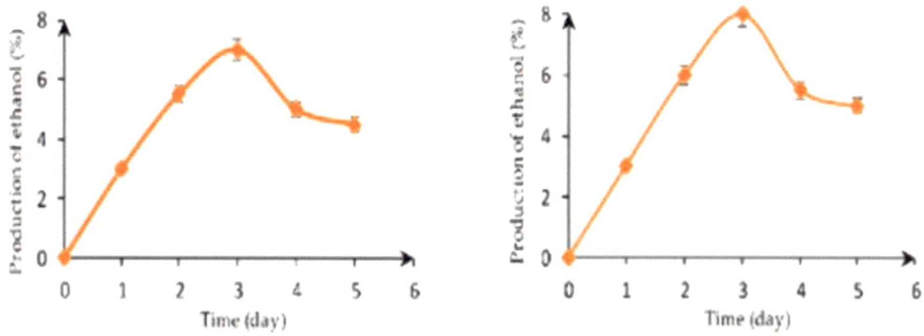

Figure 7.
The production of ethanol by S. cerevisiae in the industrial medium in (a) aerobic conditions and (b) aerobic–anaerobic conditions [68].

Inoculum concentration does not have any significant effect on the production of ethanol but the ethanol consumption rate and sugar yield [69]. When the is an increase in the number of cells from 1 × 104 to 1 × 107 cells per ml, increased ethanol production is also observed. It has been reported that when Inoculum concentration exceeds 107 and 108 cells per ml, no significant effect on the ethanol production observed [28]. At the elevated concentration of inoculum, reduction of fermentation time is observed as there is rapid cell growth.

8. Conclusion

The total results revealed the vegetables and fruits waste could be utilized for the production of bioethanol from recycled agricultural waste and management process. The discussions showed that bioethanol optimum yield is produced at pH 4, the temperature at 32°C and using 3 g/L yeast. The engine cars utilized efficiently bioethanol produced from waste rotten pineapple because it does not have high content and any dangerous elements. The principle or idea of using vegetables and fruits waste to produce bioethanol will aid in keeping the environment clean from the waste of agriculture. The process helped in overcoming to the challenges of depletion of fossil fuel with the creation of bioresearch energy. Bioethanol produced from the agricultural waste of vegetables and fruits is of good qualities with making the engine to produce less emission. Vegetables and fruits waste are good economical choice for the production of bioethanol because of its low cost and availability.

Acknowledgements

The authors wish to acknowledge the support of Center for Research Excellence and Incubation Management (CREIM), Universiti Sultan Zainal Abidin (UniSZA), Malaysia, Gong Badak, 21300 Kuala Nerus, Terengganu Darul Iman, Malaysia and KPT Grant (Project code: FRGS/1/2019/WAB01/UNISZA/02/2).

Author details

Mohammad Moneruzzaman Khandaker[1*], Umar Aliyu Abdullahi[1], Mahmoud Dogara Abdulrahman[2], Noor Afiza Badaluddin[1] and Khamsah Suryati Mohd[1]

1 School of Agriculture Science and Biotechnology, Faculty of Bioresources and Food Industry, Universiti Sultan Zainal Abidin, Besut Campus, Terengganu, Malaysia

2 Department of Biological Sciences, Faculty of Science, Kaduna State University, Kaduna, Nigeria

*Address all correspondence to: moneruzzaman@unisza.edu.my

References

[1] Ali MN, Khan Mohd M. Production of bioethanol fuel from renewable agro-based cellulosic wastes and waste newspapers. International Journal of Engineering Science and Technology, 2011; 3: 884.

[2] Saifuddin M, GohP, HoE, SaniW, Moneruzzaman KM, Nasrulhaq-Boyce A. Biodiesel production from waste cooking palm oil and environmental impact analysis. Bulgarian Journal of Agricultural Science, 2014; 20 (1): 186-192.

[3] Arapoglou D, VarzakasT,Vlyssides A and Israilides C . Ethanol production from potato peel waste (PPW). Waste Management, 2010; 30(10): 1898-1902.

[4] Hossain ABMS, FazlinyAR. Creation of alternative energy by bio-ethanol production from pineapple waste and the usage of its properties for the engine. African Journal of Microbiological Research, 2010; 4(9): 813-819.

[5] Galbe M, Zacchi G. A review of the production of ethanol from softwood.Applied Microbiology and Biotechnology,2002; 59(6): 618-628.

[6] Chin KL, H'ngPS. A real story of bioethanol from biomass: Malaysia perspective. In Biomass Now-Sustainable Growth and Use.InTech. 2013.

[7] Zabed H, Faruq G, Sahu J, Boyce A, Ganesan P. A comparative study on normal and high sugary corn genotypes for evaluating enzyme consumption during dry-grind ethanol production, Chem Eng J, 2016; 287: 691-703.

[8] Mabee WE, Saddler JN. Ethanol from Lignocellulosics: Policy options to support bioethanol production, IEA Task 39, Forest Products Biotechnology, University of British Columbia. Report T39-P2, 2005; 1-16 .

[9] Wheals AE, Bassoc LC, Alves DMG, Amorimd HV. Fuel ethanol after 25 years. Trends Biotechnol, 1999; 17(12): 482-487.

[10] Grad P. Biofuelling Brazil—an overview of the bioethanol success story in Brazil. Biofuels,2006;7(3): 56-59.

[11] Khandaker MM, Qiamuddin KB, Majrashi A, Dalorima T, Sajili MH, Hossain SABM. Bio-Ethanol Production from Fruit and Vegetable Waste by Using *Saccharomyces cerevisiae*. Bioscience Research, 2018;15(3):1703-1711.

[12] Boekhout T, Krtzman CP. Principles and methods used in yeast classification and an overview of currently accepted yeast genera. In: Wolf K., editor. Nonconventional Yeasts in Biotechnology. Springer-Verlag; Berlin: (1996); 1-99.

[13] Mohd Azhar SH, Abdulla R, Jambo SA, Marbawi H, Gansau JA, Mohd Faik AA, Rodrigues KF. Yeasts in sustainable bioethanol production: A review. Biochemistry and Biophysics Reports, 2017; 10, 52-61.

[14] Itelima J, Onwulin F, Onwulin E, Isaac O, Forji S. Bio-ethanol production from banana, plantain and pineapple peels by simultaneous saccharification and fermentation process. Int J Environ Sci Develop, 2013; 4: 2137.

[15] Jahid M, Gupta A, Sharma DK. Production of Bioethanol from Fruit Wastes (Banana, Papaya, Pineapple and Mango Peels) Under Milder Conditions. J Bioprocess Biotech 2018; 8: 327. doi:10.4172/2155-9821.1000327

[16] Somrutai, K. chemistryindustry.biz. https://www.pinterest.com/pin/, 2020; 2296132809771

[17] Sulaiman A, Othman N, Baharuddin AS, Mokhtar MN,

Tabatabaei M: Enhancing the halal food industry by utilizing food wastes to produce value-added bioproducts. Procedia Soc Behav Sci, 2014; 121:35-43.

[18] Singh A, Kuila A, Adak S, Bishai M, Banerjee R. Utilization of Vegetable Wastes for Bioenergy Generation. Agric Res, 2012; 1(3):213-222.

[19] Promon SK. Ethanol production using vegetable peels medium and the effective role of cellulolytic bacterial (*Bacillus subtilis*) pre-treatment. 2015; Bangladesh: BRAC University.

[20] Pleissner D, Lam WC, Sun Z, Lin CSK: Food waste as nutrient source in heterotrophic microalgae cultivation. Bioresource Technology, 2013; 137:139-146.

[21] Pleissner D, Kwan TH, Lin CSK: Fungal hydrolysis in submerged fermentation for food waste treatment and fermentation feedstock preparation. Bioresource Technology, 2014; 158:48-54.

[22] Wilson K, Lee AF. Bio-based chemicals from biorifi ning: carbohydrate conversion and utilisation. In: Advances in Biorefi neries, Woodhead, Waltham, MA, 2014.

[23] Zhao XQ, Zi LH, Bai FW, Lin HL, Hao XM, Yue GJ, Ho NW. Bioethanol from lignocellulosic biomass. In Biotechnology in China III: Biofuels and bioenergy, Springer, Berlin, Heidelberg, 2011; 25-51.

[24] Del Campo I, Alegria I, Zazpe M, Echeverria M, Echeverria I. Diluted Acid Hydrolysis Pretreatment of Agri-food Wastes for Bioethanol Production. International Journals of Industrial Crops and Products, 2006. 24: 214-221

[25] Supriyanto A, Lestari I, Citrasari N, Putri SS. The bioconversion of vegetable waste extract from Osowilangun Central Market Surabaya into bioethanol using *Saccharomyces cerevisiae*. IOP Conf. Series: Earth and Environmental Science 245 (2019) 012042.

[26] Bhadana B, Chauhan M. Bioethanol Production Using *Saccharomyces cerevisiae* with Different Perspectives: Substrates, Growth Variables, Inhibitor Reduction and Immobilization. Fermentation Technology, 2016; 5 (2): 1-5.

[27] Agustini, L. Efiyanti, L. Faulina, SA. Santoso, E. Isolation and characterization of cellulase-and xylanase-producing microbes isolated from tropical forests in Java and Sumatra. Int J Environ Bioeng. 2012; 3:154-167.

[28] Zabed H, Faruq G, Sahu JN, Azirun MS, Hashim RA, Boyce N. Bioethanol production from fermentable sugar juice, Sci World J, 2014; 2014:1-11.

[29] Swain MR, Krishnan C. Improved conversion of rice straw to ethanol and xylitol by combination of moderate temperature ammonia pretreatment and sequential fermentation using *Candida tropicalis*, Ind Crops Prod, 2015; 77:1039-1046.

[30] Prasad S, Singh A, Joshi H. Ethanol as an alternative fuel from agricultural, industrial and urban residues, Resour Conserv Recycl, 2007; 50:1-39.

[31] Cavka, A. Jonsson LJ. Detoxification of lignocellulosic hydrolysates using sodium borohydride, Bioresour Technol, 2013; (136): 368-376.

[32] Graf A, Koehler T. Oregon cellulose-ethanol study: an evaluation of the potential for ethanol production in Oregon using cellulose based feedstocks. Salem, Oregon, USA: Oregon Dept of Energy. 2000; 1-36.

[33] Mishima D, Tateda M, Fujita M. Comparative study on chemical

pretreatments to accelerate enzymatic hydrolysis of aquatic macrophyte biomass used in water purification processes. Bioresource Technology 2008; 97: 2166-2172.

[34] Xiang Q, Lee YY, Pettersson OP, Robert W, Torget. Heterogeneous aspects of acid hydrolysis of α-cellulose. Applied Biochemistry and Biotechnology,2004; 10: 505-514.

[35] Geng A, Xin F, Jun-yu Ip. Ethanol production from horticultural waste treated by a modified organosolve method. Bioresource Technology, 2009; 104:715-721.

[36] Devi S, Dhaka A, Singh J. Acid and alkaline hydrolysis technologies for bioethanol production: an overview. International Journal of Advanced Technology in Engineering and Science, 2016;4(6):94-106.

[37] Chandel AK, Es C, Rudravaram R. Economics and environmental impact of bioethanol production technologies: an appraisal. Biotechnol. Mol. Biol. Rev. 2007; 2:14-32

[38] Demirbas A. Biofuels sources, biofuel policy, biofuel economy and global biofuel projections. Energy Conversion and Management, 2008; 49(8): 2106-2116.

[39] Farooqi R, Sam AG. Ethanol as a transportation fuel. Centre for Applied Business Research in Energy and the Environment (CABREE) Climate Change Initiative. University of Alberta, Canada, 2004; [cited; available from: /www.business.ualberta.ca/cabreeS].

[40] Cardoso WS, Tardin FD, Tavares GP, Queiroz PV, Mota SS, Catarina M, Kasuya M, Queiroz JHD. Use of sorghum straw (*Sorghum bicolor*) for second generation ethanol production: pretreatment and enzymatic hydrolysis. Quim Nova, 2013; 36(5):623-627.

[41] Bensah EC, Mensah M. Chemical pretreatment methods for the production of cellulosic ethanol: technologies and innovations. Int J Chem Eng. 2013; doi:10.1155/2013/719607

[42] Rabelo SC, Filho RM, Costa AC. A comparison between lime and alkaline hydrogen peroxide pretreatments of sugarcane bagasse for ethanol production. Appl Biochem Biotechnol,2008; 144: 87-100.

[43] Koppram R, Tomas-Pejo E, Xiros C, Olsson L. Lignocellulosic ethanol production at high-gravity: challenges and perspectives. Trends Biotechnol 2014; (32)46-53.

[44] Canilha L, Chandel AK, dos Santos Milessi TS et al. Bioconversion of Sugarcane Biomass into Ethanol: an overview about composition, pretreatment methods, detoxification of hydrolysates, enzymatic saccharification, and ethanol fermentation. J. Biomed. Biotechnol. (2012); 1-15.

[45] Hadiyanto H, Ariyanti D, Aini AP, Pinundi DS. batch and fed-batch fermentation System on ethanol production from whey using Kluyveromyces marxianus, Int. J. Renew. Energy Dev. 2013; 2; 127-131.

[46] Thatoi H, Dash PK, Mohapatra S, Swain MR. Bioethanol production from tuber crops using fermentation technology: a review, Int. J. Sustain. Energy, 2014; 1-26.

[47] Jain A. Chaurasia SP. Bioethanol production in membrane bioreactor (MBR) System: a review, Int. J. Environ. Res. Dev. 2014; (4) 387-394.

[48] Cheng NG, Hasan M, Kumoro AC. Production of ethanol by fed-batch fermentation, Pertanika J. Sci. Technol. 2009; 17: 399-408.

[49] Dabelstein W, Reglitzky A, Schütze A, Reders K. "Automotive Fuels"

in Ullmann's Encyclopedia of Industrial Chemistry, Wiley-VCH, Weinheim, 2007; doi:10.1002/14356007.a16_719.

[50] Ullmann. Ullmann's Encyclopedia of Industrial Chemistry Ethanol 1990; 588-9.

[51] Ritslaid K, Kuut A, Olt J. 2010 State of the Art in Bioethanol Production Agron. Res. 2010; 8: 236-54.

[52] de Menezes EW, Cataluña R, Samios D, Silva RD. Addition of an azeotropic ETBE/ethanol mixture in eurosuper-type gasolines. Fuel, 2006; 85: 17-18.

[53] Demirbas, A. Estimating of structural composition of wood and non-wood biomass samples. Energy Source, 2005; 27: 761-7.

[54] Bradley D, Head RA. Engine autoignition: the relationship between octane numbers and autoignition delay times. Combust Flame, 2006; 147: 171-184

[55] Walker GM. Fuel ethanol: current production and future challenges. Journal of the Institute of Brewing, 2011; 117 (1):3-22.

[56] Jegannathan KR, Chan ES, Ravindra P. Harnessing biofuels: a global Renaissance in energy production?" Renewable and Sustainable Energy Reviews, 2009; 13 (8):2163-2168.

[57] Deenanath ED, Rumbold K, Iyuke S. The Production of Bioethanol from Cashew Apple Juice by Batch Fermentation Using *Saccharomyces cerevisiae* Y2084 and Vin13. ISRN Renewable Energy, 2013, 1-11.

[58] Lin Y, Zhang, W, Li C, Sakakibara K, Tanaka S, Kong H. Factors affecting ethanol fermentation using *Saccharomyces cerevisiae* BY4742. Biomass and Bioenergy, 2012; 47, 395-401. doi:10.1016/j.biombioe.

[59] Phisalaphong M, Srirattana N, Tanthapanichakoon W. Mathematical modeling to investigate temperature effect on kinetic parameters of ethanol fermentation. Biochem Eng J, 2006;28:36-43.

[60] McMeckin TA, Olley J, Ratkwsky DA, Ross T. Predictive microbiology: towards the interface and beyond. Int J Food Microbiol 2002; 73:395-407.

[61] Fakruddin Md, Abdul Quayum Md, Ahmed MM, and Choudhury N. (2012). Analysis of Key Factors Affecting Ethanol Production by *Saccharomyces cerevisiae* IFST 072011. Biotechnology, 2012; 11: 248-252.

[62] Gao C, Fleet GH. The effects of temperature and pH on the ethanol tolerance of the wine yeasts: *Saccharomyces cerevisiae*, Candida stellata and *Kloeckera apiculata*. J Appl Bacteriol 1988; 65:405-10

[63] Torija MJ, Rozes N, Poblet M, Guillamon JM, Mas A. Effects of fermentation temperature on the strain population of *Saccharomyces cerevisiae*. Int J Food Microbiol, 2003;80:47-53

[64] Triwahyuni E, Sudiyani Y, Abimanyu H. The effect of substrate loading on simultaneous saccharification and fermentation process for bioethanol production from oil palm empty fruit bunches Energy Procedia, 2015; 68 138-146.

[65] Hosny M, Abo-State MA, El-Temtamy SA, El-Sheikh HH. Factors Affecting Bioethanol Production from Hydrolyzed Bagasse. Int. J. Adv. Res. Biol. Sci. 2016; 3(9): 130-138.

[66] Staniszewski M, Kujawski W, Lewandowska M. Ethanol production from whey in bioreactor with co-immobilized enzyme and yeast cells followed by pervaporative recovery of product – Kinetic model predictions, J. Food Eng. 2007; 82:618-625.

[67] Wong YC, Sanggari V. Bioethanol Production from Sugarcane Bagasse using Fermentation Process. Oriental Journal of Chemistry, 2014; 30(2): 507-513. doi:10.13005/ojc/300214

[68] Darvishi F, Abolhasan, Moghaddami N. Optimization of an Industrial Medium from Molasses for Bioethanol Production Using the Taguchi Statistical Experimental-Design Method. Fermentation, 2019; 5(1): 14. doi:10.3390/fermentation5010014

[69] Laopaiboon L, Thanonkeo P, Jaisil P, Laopaiboon P. Ethanol production from sweet sorghum juice in batch and fed-batch fermentations by *Saccharomyces cerevisiae*. World J. Microbiol. Biotechnol. 2017;(23):1497-1501.

Chapter 4

Comparative Analysis of Bioethanol Production from Different Potential Biomass Sources in the Philippines

Kristel M. Gatdula, Rex B. Demafelis and Butch G. Bataller

Abstract

To pursue the continuous implementation of the bioethanol blending mandate by the Philippine Biofuels Law, part of the roadmap of the National Biofuels Board (NBB) through the Department of Energy (DOE) is to find a sustainable feedstock. This is due to the deficit in locally produced bioethanol as there is an insufficient supply of currently used feedstock, sugarcane. There are several biomasses available in the country with components viable for ethanol fermentation. Aside from sugarcane, these include sweet sorghum and cassava (first-generation), rice straw and corn stover (second-generation), and macroalgae (third-generation). Among which, sweet sorghum can be considered as the best complementary feedstock to sugarcane as its syrup can be directly fermented to produce bioethanol. Considering its maximum bioethanol potential yield of 100 L/ton for two croppings annually, a comparably low production cost of PhP 36.00/L bioethanol was estimated, competitive enough with the PhP33.43/L bioethanol from sugarcane. Aside from finding a promising feedstock, the bioethanol production volume in the country must be increased to meet the demand through either working on the optimum processing conditions to increase the capacity utilization from the current 77.9% or through installation of additional distilleries.

Keywords: bioethanol, biomass, fermentation, lignocellulosic, saccharine, starchy feedstocks

1. Introduction

The use of biofuels in the Philippines was initiated due to the oil crisis of the 1970s. However, the implementation did not immediately push through as the domestic cost of production of biofuels was higher than the cost of importing oil. Hence, to reduce the biofuels production cost and correspondingly its selling price, the Philippine Department of Agriculture (DA), as a member of the NBB, is tasked to develop a sustainable and viable feedstock for the production of biofuels. For each feedstock assessed based on its availability and accessibility, a suitable and economically competitive conversion technology is applied, usually developed through the research programs of the Philippine Council for Industry, Energy and Emerging Technology Research and Development (PCIEERD) of the Department of Science and Technology (DOST). On the other hand, for the implementing policies supporting the Philippine Biofuels program, the enactment of the Republic Act No. 9367 (RA 9367), also known as the Biofuels Act of 2006, designates specific mandates on government agencies particularly to ensure feedstock supply like sugarcane for bioethanol under the Sugar Regulatory Administration (SRA) and coconut for biodiesel under the Philippine Coconut Authority (PCA) [1].

RA 9367 aims to increase the contribution of biofuels in the country's energy mix, specifically in the transport sector, by reducing its dependence on fossil-based fuels, mitigating climate change while creating more job opportunities for the national socio-economic development. The law took effect last February 6, 2007, and after two years, 2% by volume blending of biodiesel with diesel and 5% by volume blending of bioethanol with gasoline were implemented. The monitoring of this blending mandate and the evaluation of policy execution are by the NBB, headed by the Philippine DOE. The law also encourages investments in the biofuels sector by providing incentives in the production, distribution and use of locally produced biofuels. Incentives include zero specific tax per liter of volume for biofuels component, exemption from value added tax (VAT) on the sale of raw material used in the production of biofuels, exemption from wastewater charges of all water effluents from the biofuels production in accordance with the Republic Act No. 9275 (Philippine Clean Water Act of 2004), and high priority from government financial institutions if engaged in production, storage handling and transport of biofuels feedstock, biofuels and blending with petroleum as certified by the Philippine DOE [2]. Additional incentives such as income tax holiday (ITH), duty-free importation, and low-income tax rate of 10% after the ITH may also be availed as implied by the Republic Act No. 9513 or the Renewable Energy Act of 2008. In tandem with these and with the aim to further boost the country's rural economy, the Joint Administrative Order No. 2008–1 was released providing guidelines governing biofuels feedstock production and emphasizing that feedstocks for biofuels production must only be sourced locally [3].

Currently, DOE has maintained the ethanol and biodiesel blends at 10% and 2% by volume, respectively and still unclear whether the blending targets based on the RA 9367 will be increased for the succeeding years. Until now, the insufficient feedstock supply is one of the major dilemmas in the ethanol industry. From the utilization of sugarcane, the ethanol industry players have been in transition to using molasses as source of bioethanol fuel. Molasses, as a byproduct of sugar manufacturing process from sugarcane, also has a limited supply resulting to unwarranted escalation in the price of gasoline when blended with bioethanol. Same case with the biodiesel industry, there is a fluctuation in the price of *copra*. *Copra* refers to the dried coconut kernel or meat from which the oil that is processed to biodiesel is obtained. Unlike sugarcane or molasses, there is oversupply of *copra*. Hence, the gradual upward adjustments in the biodiesel blend would not be a problem.

To provide sufficient feedstock to existing ethanol plants, the joint Oversight Committee on Energy of the Philippine Congress and Senate arrived with the interim solutions last year 2019. These include exploring the high yielding varieties of sugarcane and revisiting the available feedstocks in the country for bioethanol production. There is also a continuous conduct of research and development on feedstock sources as imposed by DOE through its Philippine Energy Plan on biofuels for the year 2020 to 2040 [4]. Some of the biomass under research and deployment for bioethanol production include sugarcane, cassava and sweet sorghum for first-generation, agricultural and forest residues such as rice straw and corn stover for second-generation, and macroalgae, commonly known as seaweeds for third-generation biofuel. Hence, this book chapter is a consolidation of information on the potential of different biomass available in the Philippines that can be utilized as feedstock for bioethanol production. A brief overview of the bioethanol market in the Philippines is also discussed which supports the need for complementary feedstocks to sugarcane for the different generations of bioethanol. The criteria for selection of a good feedstock for bioethanol, challenges that maybe encountered upon using the biomass as starting raw material, and appropriate conversion technologies are also provided.

2. Bioethanol situation in the Philippines

2.1 Bioethanol market size and trends

A joint sales report by the Chamber of Automotive Manufacturers of the Philippines, Inc. (CAMPI) and Truck Manufacturers Association (TMA) showed a boost in vehicle sales of 3.5% relative to last year 2018 [5]. The increase in purchase of cars, motorcycles, tricycles, and utility vehicles had been attributed to the increasing income of the middle class and improvements in the infrastructure. In addition, the lowering of the amortization costs of most vehicles and lower interest rate make the acquisition of these vehicles more affordable to many. This trend is projected to continue in the future. This surge in the vehicle industry sales brings in the transport sector as the most energy intensive sector contributing the 37.2% in the country's final energy demand. This corresponds to 39.9% gasoline's share out of the 96.4% bulk share of petroleum products used as the sector's primary fuel. The continuous growth in gasoline consumption brought by rapid urbanization and motorization would require larger volume of ethanol [6].

For the year 2020, a higher blend of 20% should have been imposed according to the Philippine Energy Plan 2016–2030 (**Figure 1**), with the goal also to increase the bioethanol blending up to 85% in 2030 [2, 6]. The planned increase is still under review due to several factors such as high feedstock and production cost, insufficient and competing feedstock supply, inadequate bioethanol production plant capacity, as well as importation and price regulations. To meet the country's ethanol demand, the Biofuels Act allowed the importation of bioethanol within four (4) years from the effectivity of the law and only to the extent of the shortage as may be determined by the NBB [7]. Imports were sourced from the United States of America, Australia, and South Korea [8]. This defeats the purpose of the National Biofuels Program which seeks to reduce dependence on imported oil. Even beyond the imposed period and in the current year, the bioethanol from local sources is still insufficient to meet the demand resulting to huge importation (**Figure 2**) despite the disallowance of the law.

There are twelve (12) accredited bioethanol producers with a total production capacity of 380.50 million liters since 2018 (**Figure 3**) [9, 10] and as of the end of March 2020 (**Table 1**) [11]. These refineries operate at 77.9% capacity (**Figure 3**) [9, 10]. Additional three (3) plants with a total capacity of 113 million liters will be on-stream supposedly in 2020, two are still under construction and one completed

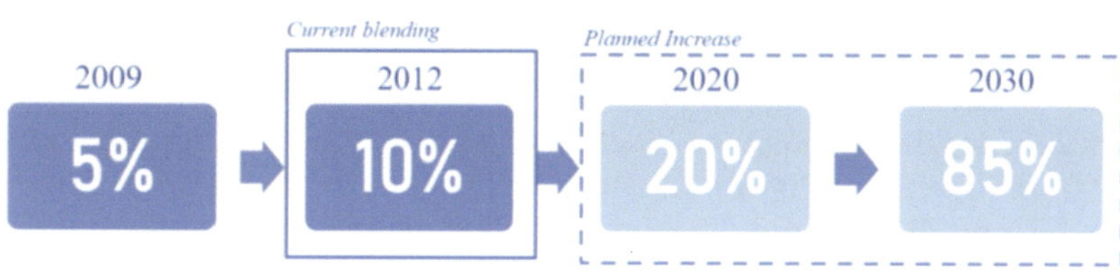

Figure 1.
Mandated bioethanol blending rate based on the Biofuels Act of 2006.

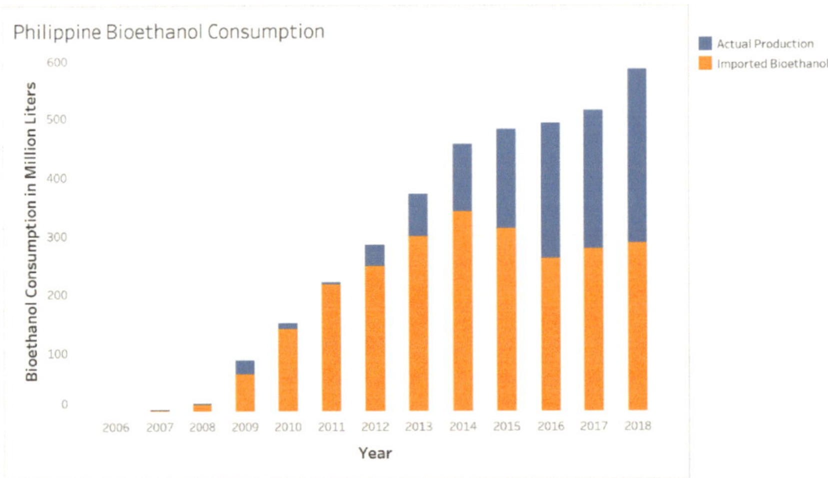

Figure 2.
Bioethanol consumption in the Philippines, 2006–2018.

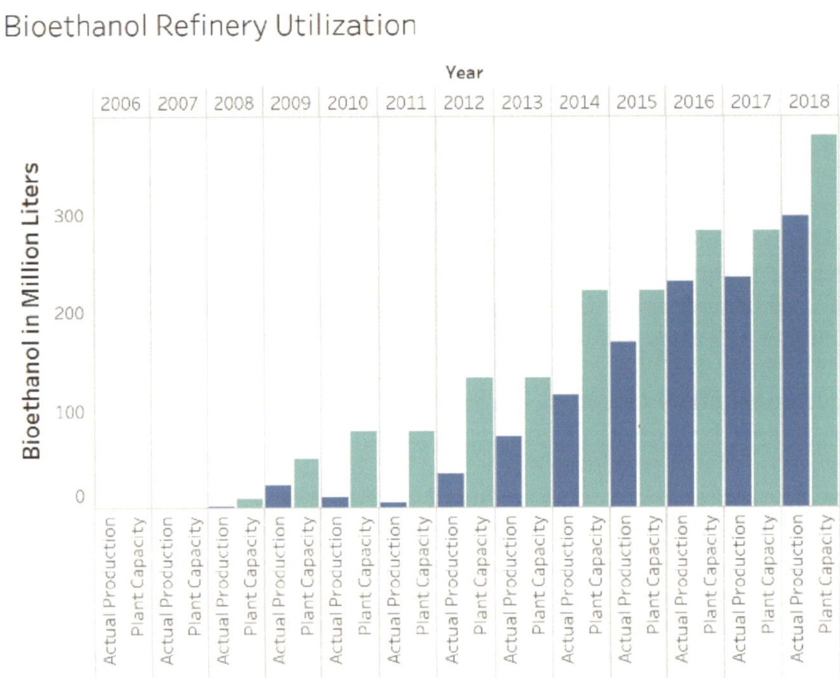

Figure 3.
Bioethanol refinery capacity utilization in the Philippines, 2006–2018.

the pilot plant test [11]. This leads to a total of 493.5 million liters production capacity, still lacking to meet the projected bioethanol demand of 847.61 million liters (supposedly at 20% blending rate) for the current year.

From 2007 to 2014, majority of bioethanol produced in the Philippines was sourced from sugarcane (**Figure 4**). Due to the SRA's mandate on prioritizing sugarcane for sugar production and the increasing demand in domestic bioethanol, the share of molasses (a by-product in sugar production) in the bioethanol feedstock mix increased in the recent years. In 2018, about 99.2% of domestic bioethanol was produced from molasses [9, 10].

Bioethanol producers	Refineries' location	Plant capacity (MLPY)
1. Absolut Distillers, Inc.	Lian, Batangas	30
2. Balayan Distillery, Inc.	Calaca, Batangas	30
3. Far East Alcohol Corp.	Apalit, Pampanga	15
4. Green Future Innovations, Inc.	San Mariano, Isabela	54
5. Progreen Agricorp, Inc. - Nasugbu	Nasugbu, Batangas	30
6. Progreen Agricorp, Inc. – Balayan	Balayan, Batangas	66
7. Kool Company, Inc.	Talisay City, Negros Occidental	30
8. Leyte Agri Corp.	Ormoc City, Leyte	9
9. Roxol Bioenergy Corporation	La Carlota City, Negros Occidental	30
10. San Carlos Bioenergy, Inc.	San Carlos City, Negros Occidental	40
11. Universal Robina Corporation	Bais City, Negros Oriental	30
12. Victorias Milling Company, Inc.	Manapla, Negros Occidental	16.5

Table 1.
List of accredited operational bioethanol producers in the Philippines as of 31 march 2020.

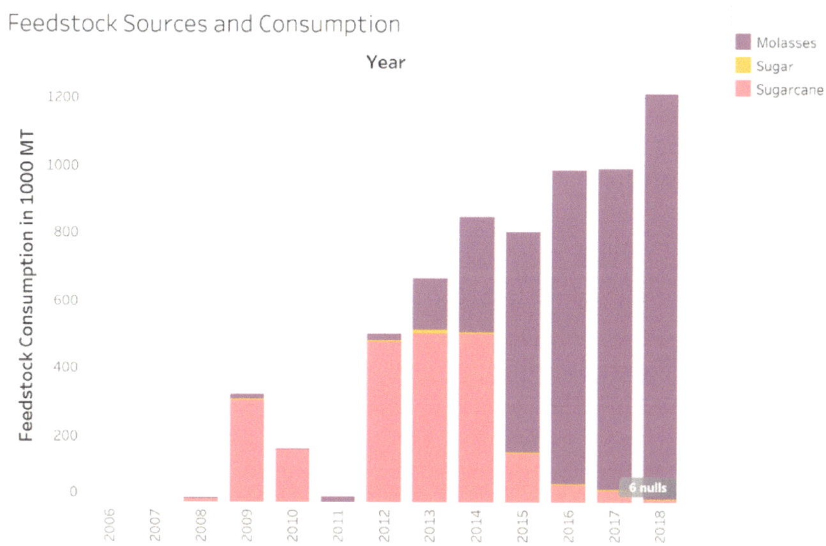

Figure 4.
Consumption of bioethanol feedstocks in the Philippines, 2006–2018.

The bioethanol price index in the Philippines is determined by the SRA bi-monthly and is based on the equivalent prices of sugarcane and molasses, the two major bioethanol feedstock sources in the country. **Figure 5** shows the monthly average prices of sugarcane, molasses, and bioethanol from 2011 to 2020 [12]. The bioethanol price shows a general slight increase during this period, which can be attributed to the increase in molasses price. The price of molasses has become more expensive than sugarcane during the period in which its share on the bioethanol feedstock mix has increased. Before, molasses was an under-utilized by-product of sugar mills in the country, but it has found value as a bioethanol feedstock after the implementation of the RA 9367. In the point of view of the sugar mills, the increase in molasses price is

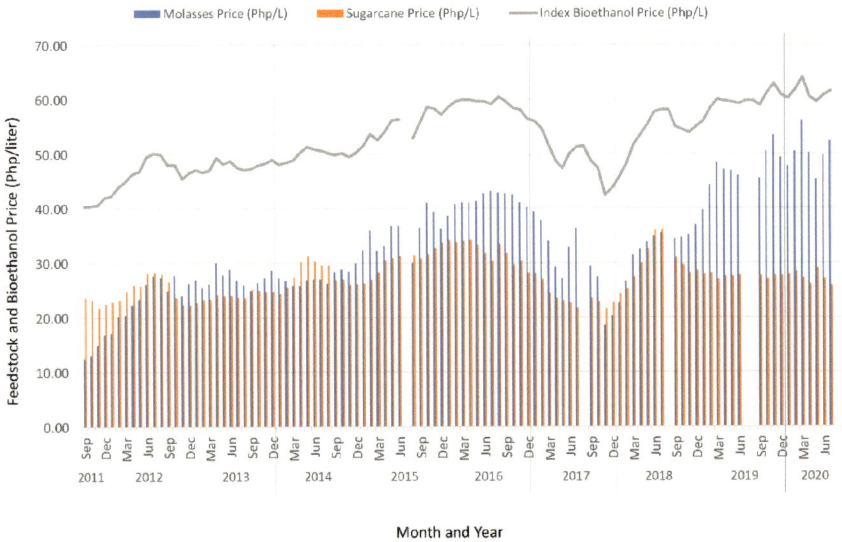

Figure 5.
Bioethanol price trend against feedstock prices, i.e. molasses and sugarcane. Gaps in the chart represent end of milling season.

advantageous due to the added income that it brings to the mills. But for the bioethanol industry, an increase in molasses price and in general, an increase in feedstock price, would mean an increase in the bioethanol selling price. Currently, bioethanol is more expensive than gasoline. Perhaps, this is one of the bottlenecks in increasing the bioethanol blend as it will drive the price of blended gasoline to increase. One of the possible ways to dampen the increase in feedstock price is to diversify feedstock sources and to increase the domestic bioethanol production capacity.

2.2 Bioethanol industry outlook

In the next 20 years, the country is still expected to heavily rely on petroleum products for fuel (**Figure 6**). With the projected increase in bioethanol blend, its demand is also expected to increase to 2616.38 million liters in 2040 [8]. More

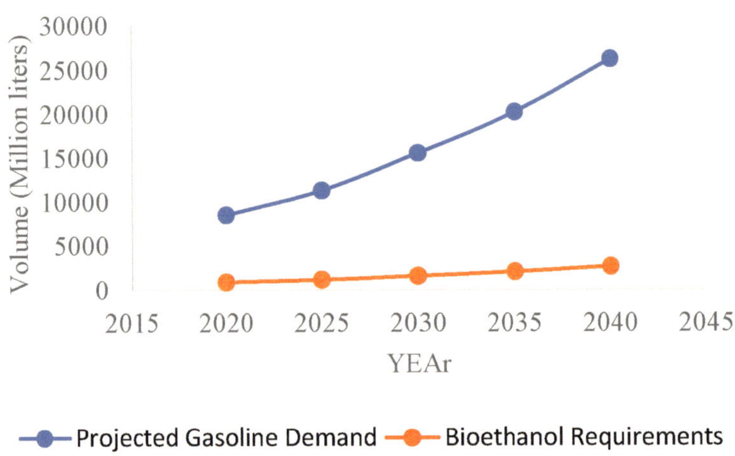

Figure 6.
Bioethanol demand outlook for 2020–2040.

Year	Gasoline requirement (MLPY)	Bioethanol requirement (MLPY)	Additional plants needed
2020	4302	860	15
2021	4381	876	1
2022	4467	893	
2023	4559	912	1
2024	4657	931	
2025	4663	933	1
2026	4757	951	
2027	4843	968	1
2028	4937	987	
2029	5005	1001	1
2030	5052	1010	

Table 2.
Additional refineries needed based on projected gasoline consumption and bioethanol requirement at 20% blending rate for the year 2020–2030.

distilleries will be needed to meet this demand as well as land area to be planted for additional bioethanol feedstock.

In the projections by the Philippine DOE, the total gasoline consumption would grow annually by 1.9%, while bioethanol at 9.7%. To meet the 2020 bioethanol demand at 20% blend without importation, the Philippines necessitates 15 additional distilleries. However, starting from 2021 and every other year, there is a need to put up five (5) more bioethanol plants to meet the blending requirements in 2030 (**Table 2**) [9].

3. Suitability assessment and challenges of producing bioethanol from biomass sources

3.1 Abundance and availability of feedstock supply

There are three generations of feedstocks that can be used as source of bioethanol and these feedstocks can be further classified based on the substrates present on their structure, suitable for ethanol conversion. One of these includes saccharine-based feedstocks or those containing sugars readily fermentable into bioethanol. Examples of these saccharine feedstocks are sugarcane and sweet sorghum. There are also land-based feedstocks that must undergo pre-processing conditions to make the substrates cellulose and hemicellulose available for ethanol fermentation. These include agricultural and forest residues. On the other hand, the third-generation biofuel feedstock such as macroalgae, commonly known as seaweed, can thrive and expand without taking the land used for agriculture. With the aim also to find a sustainable source of fuel that will not compete with food production, macroalgae's potential as feedstock was explored.

As shown in **Table 3**, sweet sorghum has the highest ethanol productivity among the potential feedstocks, based on the availability of lands for cultivation and with the assumption of maximum full biomass conversion to bioethanol. Sweet sorghum is drought tolerant and can survive in a range of environmental conditions making it a suitable complementary feedstock to sugarcane. Like sugarcane, sweet sorghum can be regenerated through ratooning. It has a shorter life cycle of four (4) months and its ratoon can be harvested for as early as three (3) months, relatively shorter compared with the 10-month crop duration of sugarcane [13]. This crop could be the solution on the issues of land availability for planting sugarcanes in the Philippines.

Rice straw is the 2nd biomass with the highest potential as source of bioethanol based on **Table 3**. If the establishment of bioethanol plants using rice straw will be realized, this biomass can contribute to about 11% of the total bioethanol demand of the country by 2030 at 85% blending rate. However, currently, there are a lot of emerging applications of rice straw which may lead to competing use of the biomass for ethanol fermentation. Rice straw can be also utilized for practices like soil improvement through carbonization and composting, fuel for power generation, mushroom production, animal feeds and as starting material to produce silica and bio-fiber [17].

In terms of maximum bioethanol potential, the seaweed *Sargassum* has the highest recorded yield with 467 L bioethanol per ton biomass. *Sargassum* is a brown macroalgae which belongs to the phylum *Phaeophyta.* It has a high content of degradable carbohydrates to bioethanol such as alginate, laminaran, mannitol, fucoidan and cellulose [18]. Most of these seaweeds are found in cold waters and

Feedstock	Biomass production yield (ton/ha)	Maximum bioethanol potential yield (L/ton)	Ethanol productivity per area (L/ha)	Potential available area (ha)	Total ethanol per available area (L)	Sources
First generation						
Sugarcane	60	65	3900	4990 (existing SRA certified areas)	19.11 M	[12]
Sweet sorghum	50	100	5000	1.21 M (idle lands suitable for sweet sorghum	6.05 B	[14]
Cassava	8	178	1424	217,978	310 M	[14]
Second generation						
Rice straw	2.6 (dry season); 3.5 (wet season)	158	964	488,500	471 M	[15]
Corn stover	5.6 (dry season; 7 (wet season)	176	2218	94,900	210 M	[15]
Third generation						
Macroalgae (*Sargassum* spp.)	33.48	467	15,635	Marine area and bioethanol productivity depend on the farm size for cultivation of seaweed		[16]

Table 3.
Bioethanol production potential of feedstock options in the Philippines.

grow in the intertidal belt and in the upper littoral region [19]. However, seaweed farming is encouraged for feedstock acquisition to avoid compromising and threatening the balance in the marine ecosystem because of uncontrolled harvesting which happened in 2014. It was only in 2018 when the Bureau of Fisheries and Aquatic Resources (BFAR) lifted the ban and allow fisherfolk to collect, sell, trade and transport *Sargassum* through Administrative Order 250–2, subject to seasonal restrictions and permit requirements [20].

3.2 Low product inhibition and reduced byproduct formation

In the Philippines, fermentation is the commonly used process in producing ethanol from various types of biomass. It is a biochemical conversion of biomass into sugars using acids or enzymes and the transformation of these sugars into ethanol and other chemicals with the aid of yeast, typically *Saccharomyces*. Parameters that have relative effects on fermentation include pH, temperature, sugar concentration, types of yeast, variation in medium constituents and in the incubation time for inoculum preparation [21]. Employing optimum fermentation conditions increases the yeast growth rate and metabolism of substrates, achieving high ethanol yield and reducing byproduct formation.

Saccharomyces is highly specific in converting glucose units to ethanol. However, no biomass contains purely glucose. Thus, low substrate concentration corresponds to low product formation. Since the total sugar in the biomass is regarded as the sum of glucose, fructose and sucrose [22] and also comprise of other types such as xylose, ribose, arabinose, sorbose, galactose, mannose, etc., high amounts of substrates remain unconverted. For some feedstocks such as starchy crops, dextrification using an enzyme α-amylase is necessary to hydrolyze first the starch into maltose (Eq. 1). This step is followed by the use of exoenzyme glucoamylase to achieve conversion of maltose into glucose (Eq. 2) for then to be converted to ethanol using *Saccharomyces*. In addition to sugars, there are also salt and sulfated polysaccharides in the biomass structure which inhibit ethanol formation. These are usually removed through chemical treatment prior to fermentation. Nutrient supplementation during fermentation is also done to enhance the activity of the yeast or to improve its substrate consumption rate.

$$2\left(C_6H_{10}O_5\right)_n\left[starch\right]+nH_2O\rightarrow nC_{12}H_{22}O_{11}\left[maltose\right] \tag{1}$$

$$C_{12}H_{22}O_{11}\left[maltose\right]+H_2O\rightarrow 2C_6H_{12}O_6\left[glucose\right] \tag{2}$$

3.3 Low ethanol production cost

Eq. (3) is the established formula for bioethanol reference price provided for producers who use molasses and sugarcane as feedstocks. The equation is a function of average feedstock price and the processing cost per liter of ethanol produced [3]. Considering these two variables in comparing the different biomass for ethanol production, sugarcane and sweet sorghum are the cheapest with PhP 33.43/L and PhP 36.00/L respectively, because both can be subjected to direct fermentation. On the other hand, the estimated production cost for seaweed-based bioethanol is around PhP 45.45/L as more pretreatment processes are required to degrade the polymers alginate, cellulose, mannitol, and laminarin into reducing sugars prior to fermentation. Highest production costs for cassava, rice straw and corn stover of PhP 64.45/L, PhP 59.39/L and 46.08 PhP/L, respectively (**Figure** 7) are due to use

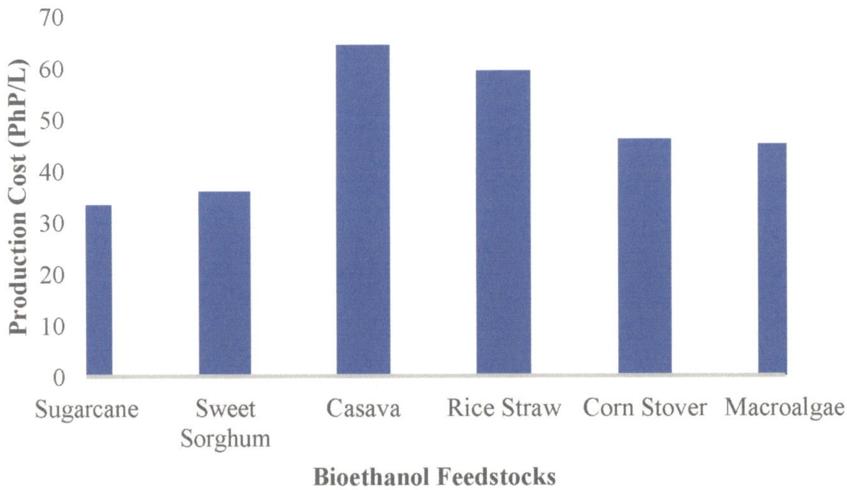

Figure 7.
Ethanol production cost per liter using various biomass.

of enzymes or additional chemicals to saccharify the starch to glucose units before fermentation.

$$EthanolPrice = AverageFeedstockPrice + ConversionCost \quad (3)$$

The low production costs of bioethanol from sweet sorghum and macroalgae lead to a competitive and comparable farmer's potential annual income from cultivating and harvesting these feedstocks, with PhP 50,000/ha per cropping for sweet sorghum and PhP 97,000/ha for *Sargassum* [14, 16].

4. Review on conversion technologies for first to third-generation bioethanol

Biothanol may be produced from sugary or saccharine feedstocks such as sugarcane and sweet sorghum, starchy feedstocks such as rice and corn, and lignocellulosic feedstocks such as agricultural and forest residues. Bioethanol may also be produced from macroalgae such as *Sargassum*.

The general process flow in producing bioethanol from these feedstocks is shown in **Figure 8**. The juice from sugarcane and sweet sorghum stalks can be directly converted to ethanol via fermentation. Starchy and lignocellulosic feedstocks need to be pretreated and saccharified prior to fermentation adding to the cost of producing ethanol. The pretreatment step is discussed further in Sections 4.1 to 4.3.

The fuel ethanol plant composed of the fermentation, distillery, and anhydrous plant, is the most critical and most complicated part of the processing plant. **Figure 9** shows a schematic diagram of the fermentation plant. It is typically composed of a series of seed culture vessels of increasing volume, pre-fermenter, and fermenters. Nutrients needed by the yeast are added to the reactors and typically composed of urea (150 ppm), di-ammonium phosphate (75 ppm). Magnesium sulfate (35 ppm) and biocide (10 ppm). Sulfuric acid is also added at 0.2 ml per 200 ml mash to maintain the pH at 4.1–4.5. The process starts with the sterilization of the substrate at 70°C for 30 minutes and then cooled. The yeast and the nutrients

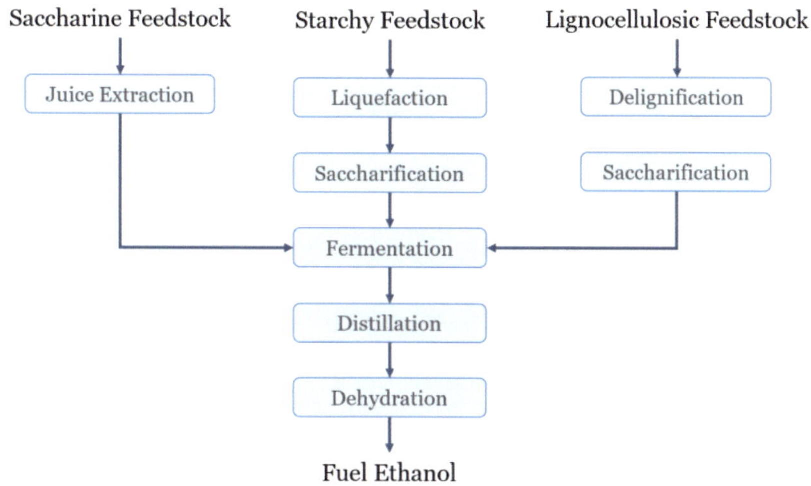

Figure 8.
General process flow of bioethanol production from different feedstocks.

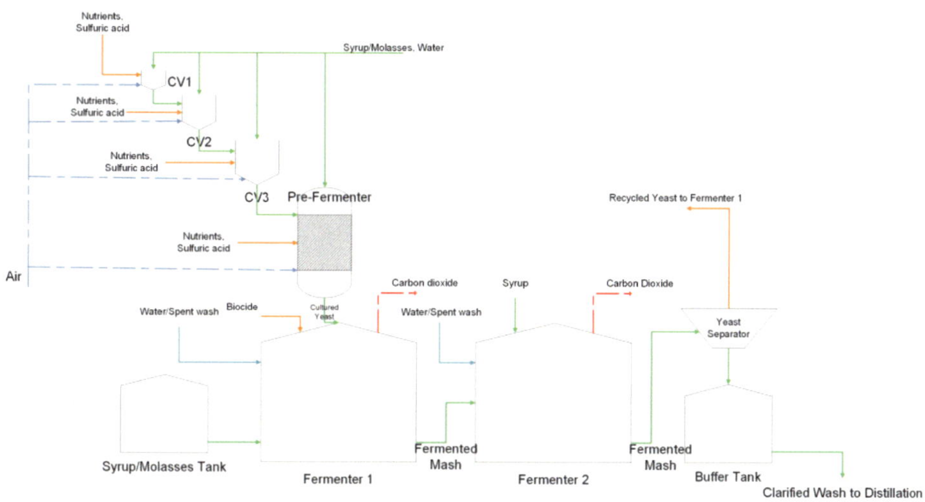

Figure 9.
Schematic diagram of fermentation in bioethanol distillery.

are then added, and fermentation is allowed for 24 hours at 32–34°C. The overall efficiency of the fermentation process is about 80–90% and the alcohol content of the clarified fermented mash is around 7–8% by volume.

The distillation plant, shown in **Figure 10**, is composed of a primary column, de-aldehyde column, and rectifier column. The primary column concentrates the alcohol content of the clarified fermented mash to about 20–30% alcohol by volume. The de-aldehyde then removes the aldehyde side-product from fermentation. The bottoms from this column is composed of 30–40% alcohol, which is fed to the rectifier column to concentrate it to 95–96% alcohol.

The dehydration plant is composed of a recovery column and two-column pressure swing adsorption (PSA) system containing molecular sieves (**Figure 11**). The recovery column removes some water from the rectified spirit, which is then superheated to 140–144°C before feeding to the PSA system. The anhydrous plant produces 99.5–00.95% ethanol with an efficiency of 99.8%.

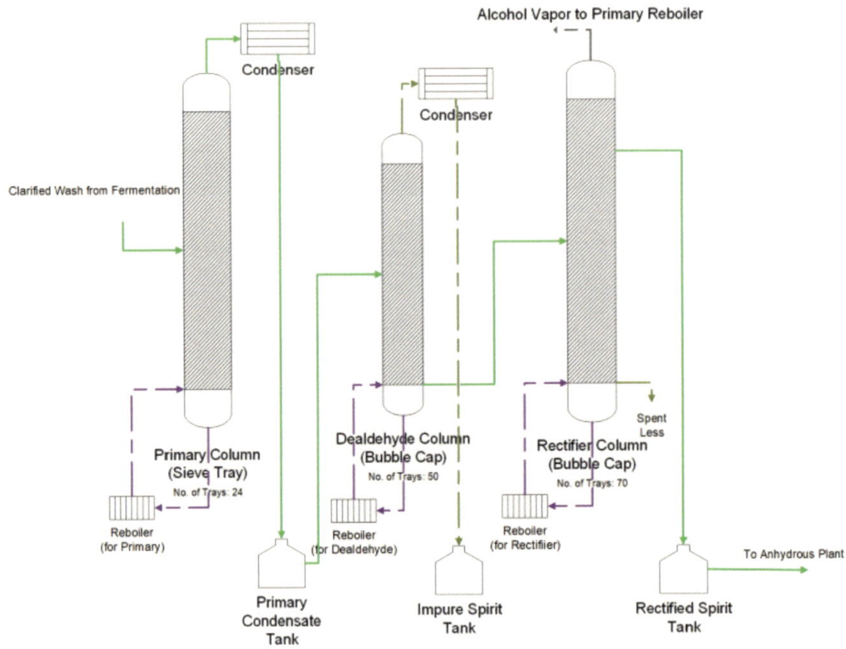

Figure 10.
Schematic diagram of distillation process in a bioethanol distillery.

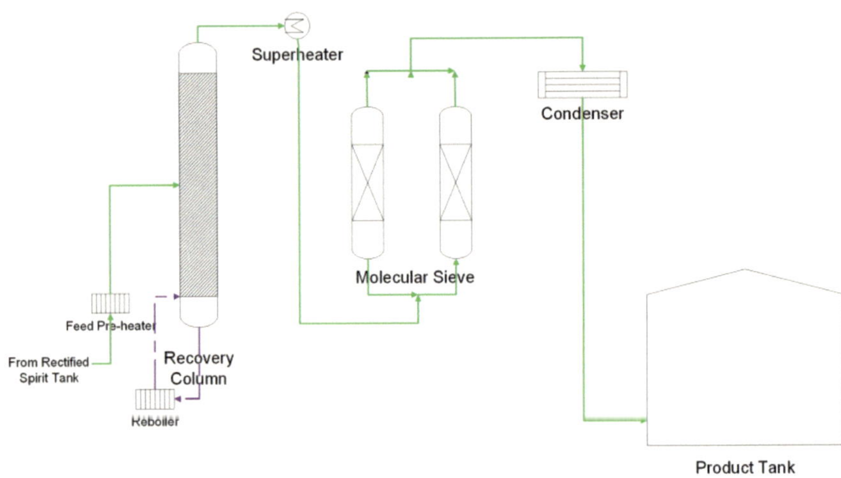

Figure 11.
Schematic diagram of dehydration by pressure-swing adsorption.

4.1 First-generation bioethanol production process

4.1.1 Sugarcane and sweet sorghum (saccharine feedstock)

Sugarcane is a grass grown in tropical and subtropical regions propagated for its sugar. In the Philippines, the sugarcane stalk yield of 60 metric ton per hectare can yield about 65 liters ethanol per metric ton per year. A 7692-hectare sugarcane plantation is needed to supply the feedstock requirement of a distillery with a 30-million-liter ethanol capacity.

On the other hand, sweet sorghum [*Sorghum bicolor* (L.) Moench] is a drought-resistant crop with high agronomic yield and sugar content of stalks. It is considered an alternative to sugarcane as a bioethanol feedstock. It can be harvested after 115 to 120 days, which makes it amenable for multiple cropping through ratooning. Stalk yield can reach 50 metric ton per hectare with an ethanol potential of 100 liters per metric ton per year. A 6000-hectare land planted to sweet sorghum will be needed to support an ethanol distillery producing 30 million liters of ethanol per year [23].

Juice extraction and syrup production from sweet sorghum requires the same operations and equipment in sugarcane processing, which includes cane handling and preparation, milling, juice heating, clarification, and evaporation.

4.1.2 Cassava (starchy feedstock)

Cassava is considered the cheapest among the major starch-based feedstock for ethanol production due to its high starch content (about 74% by weight) and starch-to-sugar conversion ratio. It can be harvested year-round in areas with evenly distributed rainfall. The crop is relatively typhoon and drought-resistant and requires minimum crop maintenance. Cassava tubers can also be chipped, dried, and stored for utilization during periods of lean supply [24].

During the period of April to June 2020, production of cassava from 110,780.04 hectares planted area was recorded at 722.82 thousand metric tons [25]. This is equivalent to a national average yield of 6.52 MT/ha. In ideal production areas for cassava, the yield can reach up to more than 20 metric tons per hectare. Ideal plantation sites are characterized by plain to rolling areas with even rainfall distribution throughout the year and soil types ranging from loam, sandy loam, fine sandy loam and silty clay loam [24].

Based on a distillery plant's capacity of 30 million liters per year, at a conservative estimate of 180 L ethanol per metric ton cassava, an estimated 8333 hectares of cassava plantation would be required to supply the feedstock for the plant's optimal operation [24].

Before processing to ethanol, the freshly harvested cassava tubers should be transformed first into cassava flour through washing, cleaning, peeling, chipping, drying, storage, and milling. The flour is then mixed with water to make a slurry, which is then gelatinized with steam, liquefied with alpha-amylase, and saccharified with glucoamylase.

Liquefaction consists of gelatinization and dextrinization. Gelatinization is the dissolution of starch into a mash by steam cooking. Gelatinization can be done in a steam jet cooker at 120°C at a residence time of only a few seconds. This step requires steam at 138°C and 2 bar pressure. The gelatinized starch then passes through a holding coil at 120°C for 1 minute and a flash tank before undergoing dextrinization. Dextrinization is the breakdown of the gelatinized starch into smaller fragments by means of alpha- or beta-amylase or dilute acid. This step results to the reduction of the solution's viscosity and the production of malto-dextrins. Dextrinization must be done immediately without allowing the solution to cool to prevent recrystallization of starch. The last pretreatment step prior to fermentation is saccharification, which can be accomplished by addition of glucoamylase and thin slops to the liquefied mash. This process is undertaken at 60°C and pH 4.5 for 30–60 minutes. The saccharification reaction is particularly fast up to 70% dextrose; as the 95% dextrose is reached, it starts slowing down [24].

4.2 Second-generation bioethanol production process

Second-generation feedstocks for bioethanol production include non-food sources such as agricultural and forest residues and grasses. They are also called lignocellulosic feedstocks since these materials are composed of lignin, cellulose, and hemicellulose. Lignin cross-links with other cell wall component making the hydrolysis of cellulose and hemicellulose to their component monomeric sugars difficult. Delignification and pretreatment are therefore required to make the cellulose and hemicellulose fraction amenable to hydrolysis and fermentation.

Pretreatment can be done physically, chemically, and biologically. Physical pretreatment involves the disruption of the lignocellulose structure by physical and mechanical means to increase the surface area of the biomass and provide access to cellulases upon hydrolysis. Physical pretreatment methods include uncatalyzed steam explosion, liquid hot water, mechanical comminution, and high energy radiation. Chemical pretreatment involves the use of chemicals to delignify the biomass and sometimes to dissolve the hemicellulose fraction to enhance enzymatic digestibility of the cellulose. This pretreatment method includes hydrolysis via concentrated acid and dilute acid, alkaline pretreatment, ammonia fiber/freeze explosion (AFEX), organosolv, pH-controlled liquid hot water, and ion liquids (ILs) pretreatment. Biological pretreatment used wood-degrading microorganisms to modify the chemical composition and/or structure of lignocellulosic biomass and make it more suitable to enzyme digestion [14].

Rice straw and corn stover are examples of lignocellulosic materials. They are agricultural wastes; thus, their cost is low if they are to be used as feedstock for bioethanol. However, the high cost of pretreatment makes them unattractive for bioethanol production. For every ton of rice produced, a ton of rice straw is also produced. At 20% moisture content, rice straw's ethanol potential is about 158 L per metric ton. At a yield of 2.6 metric ton of rice per hectare during the dry season, about 50,215 hectares of land planted to rice is needed to support an ethanol plant with a capacity of 30 million liters per year. For corn stover, 20,877 hectares of land planted to corn is needed to provide enough feedstock to an ethanol plant producing 30 million liters per year considering that 3.92 metric ton corn stover can be obtained per hectare and 176 liters ethanol can be produced per metric ton of corn stover [26].

4.3 Third-generation bioethanol production process

Over 80 species of *Sargassum* are found in the Philippines, widely distributed in the islands of Luzon, Visayas and Mindanao. It has several local names like *Aragan*, *Boto-boto*, *Lusay-lusay* and *Samo*. In the old days, *Sargassum* is used as a wrapper to maintain the freshness of fish and other sea commodities. It also functions as animal feed, a healthy beverage, fertilizer and other agricultural uses. Nowadays, the use of these species has been lessened because of modern technologies and concerns on preservation of marine biodiversity [27].

The most notable characteristic of brown macroalgae, *Sargassum* in particular, is the absence of lignin in its cell walls [18]. Lignin, a component of biomass which holds typically the celluloses, is a protective material that provides rigidity to the biomass and the resistance from microbial attack [28]. In **Figure 12**, even though there is no need for delignification process, series of pretreatment methods are necessary to convert the target polymeric constituent to bioethanol.

The seaweed shall undergo physical treatment such as drying and milling for the biomass to be more susceptible to enzyme or chemical. Mannitol can be precipitated through hot (i.e. boiling) water extraction at 96.1% efficiency. Alginic acid, on the other hand, is produced by mixing the seaweed with 4% w/w sodium bicarbonate solution. The supernatant that will be collected after treatment should be stored at 4°C prior to precipitation of alginic acid by adding sulfuric acid. Laminarin can be extracted through sequential hydrochloric acid and ethanol treatment at a working temperature of 70°C. Hydrolysis of cellulose and laminarin can be carried out using

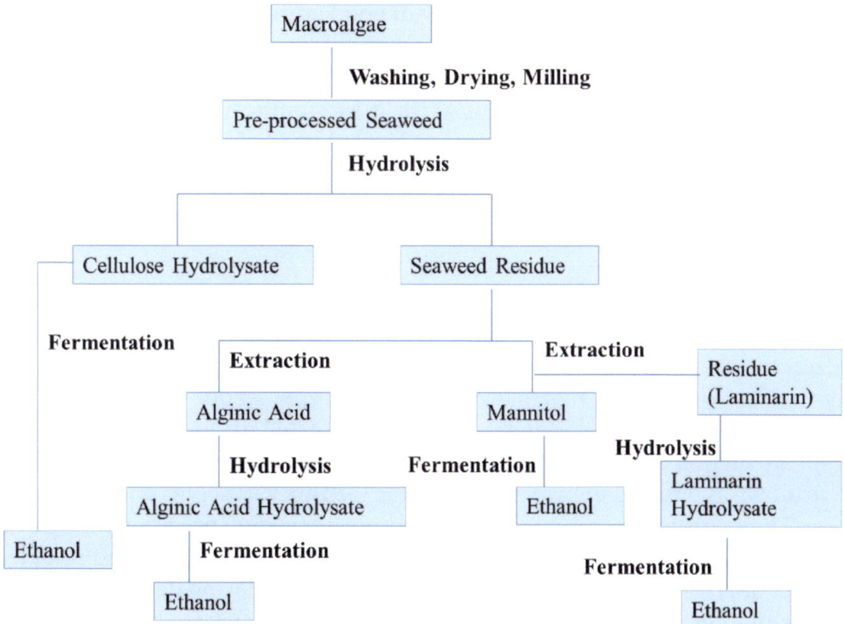

Figure 12.
Process flow of bioethanol production from Sargassum.

the enzyme *cellulase* at 50°C, while the alginic acid using the enzyme lyase at 37°C. Mannitol as a sugar alcohol can be directly fermented into bioethanol [29].

Upon fermentation and purification, a maximum ethanol yield of 467 L per ton dried seaweed could be attained from *Sargassum* with the following composition (w/w): 21% laminarin, 5% mannitol, 32.65% alginic acid and 6.20% cellulose [16]. To produce 30 million liters of bioethanol, which is the typical capacity of distilleries established in the Philippines, a cultivation area of 1918 ha of *Sargassum* is needed [29]. The seaweed composition dictates the resulting ethanol yield. The variation in composition depends on the maturity of the seaweeds. The maturity can be assessed based on the length of the plant from the holdfast to the tip of the longest shoot. The average of increase in lengths was referred as the periodic mean thallus length. The maturity is also influenced by the habitat as the rocky coralline environment of the sea was reported to be favorable for *Sargassum* growth [30].

5. Recommendations on enhancing biomass use for bioethanol production

There are two major gaps in the Philippine bioethanol industry: insufficient feedstock availability and the high domestic ethanol price due to the inefficiency of the process. In selection of complementary feedstock to sugarcane, the potential of crops and their residues should be screened out based on three sustainability criteria: social, economic, and environmental aspects to ensure that the implementation is attainable in the long run. In coordination with an agriculturist, best farm practices and agronomic conditions when the crops should be harvested considering its substrate content (i.e. through sugar analysis) should be recommended. On the processing side, it must be ensured to reduce the byproduct generation by optimizing the fermentation efficiency or the microbial activity and/or improving the pretreatment for lignocellulosic and starchy feedstocks. For agricultural crops as starting raw materials, yield intensification or breeding of good crop varieties are recommended for sufficient feedstock supply. Identification of potential cultivation areas for expansion should also be initiated to continuously sustain the bioethanol blending mandate. Policy support is also one of the most significant on these renewable energy efforts, through concerned agencies headed by the National Biofuels Board.

6. Conclusions

The enactment of Philippines RA 9367, also known as the Biofuels Act of 2006 makes the country the pioneer in biofuels blending in the Southeast Asia. The law mandates the blending of bioethanol to all gasoline sold in the country which is currently at 10%. However, the bioethanol demand was not met due to insufficient supply of locally sourced bioethanol. Hence, the country resorted to importation to sustain the implementation of RA 9367. DOE which serves as the lead agency of NBB developed roadmap and issued policies to address the feedstock concerns on bioethanol production. These include the proposal for an improved breeding program for sugarcane which is currently used as feedstock for bioethanol. Another goal is to find complementary or alternative feedstocks through extensive research and developments.

In this chapter, the criteria for selecting an ideal feedstock for bioethanol production was based on the abundance and availability of feedstock supply, low product inhibition and reduced byproduct formation, and low ethanol production cost. Upon considering the maximum bioethanol potential and available area for harvesting, the following biomass is recommended for producing bioethanol: sweet sorghum (first-generation), rice straw (second-generation), and macroalgae (third-generation). Among these three, sweet sorghum has the lowest ethanol production cost comparable to sugarcane. Hence, it can be regarded as good complementary feedstock to sugarcane for ethanol production.

The potential of other biomass as source of bioethanol can still be enhanced by applying the appropriate conversion technologies and by working on the optimum conditions of the five (5) major processes: Pretreatment, Saccharification, Fermentation, Distillation and Dehydration. The pre-processing methods of the biomass vary depending on its compositional structure (i.e. saccharine, starchy or lignocellulosic). Aside from the aim of increasing the process efficiency, technical support must also be provided to farmers and fisherfolks on sustainable management of feedstocks.

Acknowledgements

Results presented in this book chapter are from researches supported by the Philippine DA-BAR, DOE, National Economic Development Authority (NEDA), Philippine Agricultural Development and Commercial Corporation (PADCC), and Philippine National Oil Company – Alternative Fuels Corporation (PNOC – PAFC). Authors would also like to thank all involved stakeholders that made the research projects successful, including the farmers, key industry players and researchers.

Conflict of interest

The authors declare that they have no known competing financial interests or affiliations that could have appeared to influence the work reported in this book chapter.

Author details

Kristel M. Gatdula*, Rex B. Demafelis and Butch G. Bataller
Department of Chemical Engineering, College of Engineering and Agro-Industrial Technology, University of the Philippines Los Baños, Los Baños, Laguna, Philippines

*Address all correspondence to: kmgatdula@up.edu.ph

References

[1] Morgera E, Kulovesi K, Gobena A. Country Case Studies: Philippines. In: Case Studies on Bioenergy Policy and Law: Options for Sustainability. Rome, Italy: Food and Agriculture Organization (FAO) of the United Nations; 2009. p. 219-247. Available from: http://www.fao.org/3/a-i1285e.pdf [Accessed: 2020-09-11]

[2] Official Gazette of the Philippine Government. Republic Act No. 9367: Biofuels Act of 2006. Philippines: Republic of the Philippines; 2007. Available from: https://www.officialgazette.gov.ph/2007/01/12/republic-act-no-9367/ [Accessed: 2020-09-11]

[3] De Guzman RB. DOE and NBB Programs and Initiatives. In: Sweet Sorghum Stakeholders Workshop; 21 April 2016; Lian, Batangas, Philippines: Department of Energy.

[4] Velasco MM. Rising Feedstock Costs May Force Shutdown of Bioethanol Plants. Philippines: Manila Bulletin; 2019. Available from: https://mb.com.ph/2019/07/16/rising-feedstock-costs-may-force-shutdown-of-bioethanol-plants/ [Accessed: 2020-09-11]

[5] Magkilat BC. Autosales Register 3.5% Growth in 2019. Philippines: Manila Bulletin, 2020. Available from: https://mb.com.ph/2020/01/14/auto-sales-register-3-5-growth-in-2019/ [Accessed: 2020-09-11]

[6] Department of Energy Philippine Energy Plan 2016-2030 [Internet]. 2016. Available from: https://www.doe.gov.ph/sites/default/files/pdf/pep/2016-2030_pep.pdf [Accessed: 2020-09-11]

[7] Department of Energy Department Circular No. 2015-06 - 0005 [Internet]. 2015. Available from: https://www.doe.gov.ph/sites/default/files/pdf/issuances/dc_2015-06-0005.pdf [Accessed: 2020-09-11]

[8] De Guzman RB. Status Updates of the Philippine Bioenergy Industry. In: GBEP Sustainability indicators Workshop; 24 June 2019; Pasay City, Philippines: Department of Energy.

[9] USDA Foreign Agricultural Service. Global Agricultural Information Network Report: Philippines. In: Philippine Biofuels Situation and Outlook. Manila, Philippines; 2014. Available from: https://apps.fas.usda.gov/newgainapi/api/report/downloadreportbyfilename?filename=Biofuels%20Annual_Manila_Philippines_10-22-2014.pdf [Accessed: 2020-09-11]

[10] USDA Foreign Agricultural Service. Global Agricultural Information Network Report: Philippines. In: Philippine Biofuels Situation and Outlook. Manila, Philippines; 2019. Available from: https://apps.fas.usda.gov/newgainapi/api/Report/DownloadReportByFileName?fileName=Biofuels%20Annual_Manila_Philippines_10-28-2019 [Accessed: 2020-09-11]

[11] Department of Energy List of Accredited Bioethanol Producers as of 31 March 2020 [Internet]. 2020. Available from: https://www.doe.gov.ph/sites/default/files/pdf/renewable_energy/list-of-accredited-bioethanol-producers-as-of-2020-03-31.pdf [Accessed: 2020-09-11]

[12] Sugar Regulatory Administration [Internet]. 2020. Available from: http://www.sra.gov.ph/ [Accessed: 2020-09-11]

[13] Demafelis RB, Angeles DE, Dizon LSH, Gatdula KM, Pector AA, Rivera HFR. Sweet Sorghum

as Feedstock: A Milestone for Bioethanol Fuel. Philippines: Department of Agriculture – Bureau of Agricultural Research and University of the Philippines Los Baños; 2018. 150 p.

[14] Rañola RF, Movillon JL, Demafelis RB. The Philippine Biofuels Industry. Philippines: Department of Agriculture – Bureau of Agricultural Research and University of the Philippines Los Baños; 2015. 240 p.

[15] Elepaño AR, Demafelis RB, Dorado MA, Bataller BG, Badayos RB, Pampolina NM, Baticados EJN, Gatdula KM. Potential Bioenergy Production from Major Agricultural Residues in the Philippines. Philippine Journal of Crop Science. 2015; 40:49-61.

[16] UPLB Biofuels Research Team. Technology Piloting of Macroalgae as Feedstock [Research Project Final Report]. Philippines: University of the Philippines Los Baños and Department of Agriculture – Bureau of Agricultural Research; 2013.

[17] International Rice Research Institute (IRRI) Rice Straw Management [Internet]. 2020. Available from: https://www.irri.org/rice-straw-management [Accessed: 2020-09-11]

[18] Horn S, Ostgaard K. Bioenergy from Brown Seaweeds [Graduate Thesis]. Norway: Norwegian University of Science and Technology; 2000.

[19] Valencia JMT, Demafelis RB, Borines MG, Gatdula KM. Bioethanol Potential of Brown Macroalgae *Sargassum* spp. Philippine Journal of Crop Science. 2015; 40(2):1-11.

[20] Ignacio RJN. Fisheries Bureau Partly Lifts Ban on Collecting Seaweed. Philippines: BusinessWorld; 2018. Available from: https://www.bworldonline.com/fisheries-bureau-partly-lifts-ban-on-collecting-seaweed/ [Accessed: 2020-09-11]

[21] Pramanik K. Parametric Studies on Batch Alcohol Fermentation using *Saccharomyces* Yeast Extracted from Toddy [Thesis]. Regional Engineering College: Department of Chemical Engineering; 2003.

[22] Sugars: Fructose/Glucose/Sucrose [Internet]. 2020. Available from: https://www.fera.co.uk/sugars-fructose-glucose-sucrose.html [Accessed: 2020-09-11]

[23] UPLB Biofuels Research Team. Supplementary Research on Large Scale Production of Quality Sweet Sorghum Syrup using the Facilities of Organic Producers in the Island of Negros Multi-Purpose Cooperative (OPTION-MPC) [Research Project Final Report]. Philippines: University of the Philippines Los Baños and Department of Agriculture – Bureau of Agricultural Research; 2012.

[24] UPLB Biofuels Research Team. Feasibility Study of Bioethanol production from Cassava in the Philippines [Research Project Final Report]. Philippines: University of the Philippines Los Baños and Department of Agriculture – Bureau of Agricultural Research; 2008.

[25] Philippine Statistics Authority. Quarterly Bulletin: Major Vegetables and Root Crops [Internet]. 2020. Available from: https://psa.gov.ph/sites/default/files/Publication-Bulletin-MVRS-Q2-2020_0.pdf [Accessed: 2020-09-11]

[26] UPLB Biofuels Research Team. Establishment of Lignocellulosic Feedstock Databank and a Single Agriculture and Forestry Bioenergy Network [Research Project Final Report – Year 1]. Philippines: University of the

Philippines Los Baños and Philippine Agriculture Development and Commercial Corporation; 2011.

[27] Montano M, Rodrigueza M, Balitaan R. Ethnobotany of *Sargassum* spp. in the Philippines. Coast Marine Science. 2006; 30(1):222-225.

[28] Moller R, Haek S, Pauly M. EPOBIO Project: Cell Wall Saccharification. United Kingdom: CPL Press; 2006.

[29] UPLB Biofuels Research Team. Technology Piloting of Macroalgae as Bioethanol Feedstock [Research Project Final Report]. Philippines: University of the Philippines Los Baños and Department of Agriculture – Bureau of Agricultural Research; 2013.

[30] Ortiz AT, Trono GCJ. Growth and Reproductive Pattern of Intertidal and Subtidal *Sargassum (Sagassaceae, Paheophyta)* populations in Bolinao, Pangasinan [Research Project Final Report]. Philippines: University of the Philippines Diliman Marine Science Institute; 2000.

Chapter 5

An Attempt in Blending Higher Volume of Ethanol with Diesel for Replacing the Neat Diesel to Fuel Compression Ignition Engines

Prabakaran Balasubramanian

Abstract

Alcohols are renewable in nature and can be manufactured from biomass. Butanol a higher alcohol, can be utilized as co-solvent to prevent the phase separation of diesel-ethanol blends as per the previous researches. This experimentation has been conducted with the blends of diesel-ethanol with various proportions of n-butanol followed by the solubility test in the temperature range of 5–25°C. The results indicate that 45% of ethanol can be blended with diesel by the assistance of 10% of n-butanol to make the final blend stable up to a temperature of 5°C for 20 days, which met the requirements of the essential properties (ASTM). Existing diesel engine has been modified as per the optimal level of parameters such as intake air temperature (IAT), fuel injection timing (FIT), nozzle opening pressure (NOP) and compression ratio (CR) obtained using Taghuchi method of L_9 orthogonal array. Arrived out parameters are 75°C of IAT, 29° before top dead centre of FIT, 210 bar of NOP and 19: 1 of compression ratio. The implementation of these parameters in diesel engine and fueling with diesel-ethanol butanol blend containing 45% ethanol produced closer performance and emissions characteristics compared to that of diesel. However, the emissions of smoke, hydrocarbon and carbon monoxide produced by the optimal blend are found to be marginally higher compared to that of diesel. These can be ratified by the introduction of after treatment systems modifications.

Keywords: bioethanol, biobutanol, Taghuchi method, low temperature

1. Introduction

Renewable sources are the major available resource to replace the dependency of diesel fuel to internal combustion engines (CI) engines. This motivates the researchers to rapid up research in finding out a replacement to fossil by alcohols or biodiesels from various oils to blend with diesel, which leads in low dependency of diesel alone to fuel CI engines. Alcohols are better than biodiesels, as most researchers reported that the higher the volume of biodiesel in diesel-biodiesel blends the higher oxides of nitrogen emissions (NO_X). The utilization of biodiesel from edible resources will dictate limited usage of edible oils as food material. Out of the alcohols ethanol [1] can be blended into diesel for fueling in diesel engine which can be manufactured from biomass. The author utilized ethanol into diesel engine in a dual fuel mode up to 80%. Author concluded that although 80% blending of ethanol is possible for blending, the increase in the ethanol content increased the ignition delay and decreased the thermal efficiency. This also resulted in misfire for the blend containing higher volume of ethanol (higher than 30%) The limitations in using biodiesel [2] as fuel was stated by the author and recommended for low volume of biodiesel (up to 20%) along with diesel in diesel engine. The author also stated that there was a significant decrease in power by the utilization of biodiesel into CI engine. This tuned the researchers to the focus on fueling the CI engines with diesohol instead of biodiesel-diesel blends. The utilization of methanol is not found attractive as this is meant as poisonous. This paved a way to utilize diesel-ethanol to fuel CI engine. Ethanol was started its attempt as fuel for CI engines from 1980s onwards. Previous researches with respect [3] to diesel-methanol, diesel-ethanol and diesel-ethanol on the solubility and performance was compiled by the author. **Table 1** shows the standard properties of diesel, ethanol and butanol [4]. From the table the research octane number of ethanol is very much higher, this will lead to higher rate of combustion and hence, ethanol has been chosen.

The author concluded that further research on the utilization of higher volume of ethyl alcohol in ethanol-diesel and higher volume of butanol in butanol-diesel blends are further progressed in low temperature analysis. Speed of the flame in incylinder [4] of CI engine using ethanol-diesel (containing 5% ethanol) was compared with diesel. The author stated that the flame spread speed was found decreased monotonously at the initial phase and remains unchanged after certain height of ullage. This can be attributed to efficiency in combustion and the oxidizing rate when fueled with ethanol-diesel blends. A study was conducted by fueling 20% of ethanol along with Jatropha methyl ester and diesel blends on the evaporation characteristics when fueled in CI engine. It was reported that the liquid penetration of the ethanol blended biodiesel-diesel blends and the vapor penetration were found matching with that of diesel. The improvement in the evaporation rate of the fuel blend was due to the higher heat of vaporization of ethanol in the blend and the higher boiling point of the biodiesel in the blend. Diesohol containing up to 19% of ethyl alcohol was studied [5] for the essential properties such as cetane number, calorific value and flash point. The author reported that the properties are found to be closer with respect to that of diesel fuel and suitable to fuel CI engine. The author also studied the characteristics of CI engine when fueled with this blend. The report indicated that a significant increase in BTE, decrease in the emissions

S.No.	Property	Diesel	Ethanol	Butanol
1.	Density (kg/m^3)	829	785	809
2.	Kinematic Viscosity (mm^2/s)	4.04	1.07	2.6
3.	Calorific Value (MJ/kg)	42.8	26.9	33.1
4.	Heat of Vaporization (MJ/kg)	—	0.92	0.43
5.	Flammability Limits, volume (%)	—	19	11.2
6.	Flash Point (°C)	74	13	35
7.	Cetane Number	50	8	25
8.	Research Octane Number	—	129	96
9.	Energy Density (MJ/L)	—	19.6	29

Table 1.
Properties of fuels standard.

and exhaust temperature by utilizing ethanol-diesel blend in CI engine. Cyclic irregularities of diesel-ethanol [6] and diesel- butanol blends were compared when fueled in CI engine as fuel to replace diesel. The author stated that the cyclic variations produced by diesel-ethanol blends were found to be a bit stronger compared to those produced by diesel-butanol blends. The author stated that the reason for this activity was by the oxygen possessed by ethanol. Most researchers attempted diesel-ethanol blends as fuel; however, attempts are limited for the fuel blend [7–10] possessing higher volume of ethanol and for low temperatures. Hence this experimental study considers the objective as utilizing higher volume of ethanol under low temperature up to 5°C with the assistance of n-butanol as co-solvent.

2. Materials and methods

2.1 Fuels used and preparation of blends

Diesel used in this study is Bharat Stage VI low sulfur diesel procured from market. Ethanol is procured from bioethanol producer who produces bioethanol [11] from waste vegetables cut wastes. These wastes are generally not utilized properly and thrown into garbage and causing land pollution to a greater extent. Butanol is also procured from a bulk manufacturer who produces butanol from food [12] wastages. This is the novelty in this study. To start with biobutanol has been blended in proportions (**Table 1**) ranging from 0–10% in increments of 1% and kept separately. **Table 2** lists the different proportions of diesel, ethanol and butanol.

Percentage of butanol	Fuels in percentage by volume										
1	D	94	89	84	79	74	69	64	59	54	49
	E	5	10	15	20	25	30	35	40	45	50
2	D	93	88	83	78	73	68	63	58	53	48
	E	5	10	15	20	25	30	35	40	45	50
3	D	92	87	82	77	72	67	62	57	52	47
	E	5	10	15	20	25	30	35	40	45	50
4	D	91	86	81	76	71	66	61	56	51	46
	E	5	10	15	20	25	30	35	40	45	50
5	D	90	85	80	75	70	65	60	55	50	45
	E	5	10	15	20	25	30	35	40	45	50
6	D	89	84	79	74	69	64	59	54	49	44
	E	5	10	15	20	25	30	35	40	45	50
7	D	88	83	78	73	68	63	58	53	48	43
	E	5	10	15	20	25	30	35	40	45	50
8	D	87	82	77	72	67	62	57	52	47	42
	E	5	10	15	20	25	30	35	40	45	50
9	D	86	81	76	71	66	61	56	51	46	41
	E	5	10	15	20	25	30	35	40	45	50

Percentage of butanol	Fuels in percentage by volume										
10	D	85	80	75	70	65	60	55	50	45	40
	E	5	10	15	20	25	30	35	40	45	50

D-Diesel, E- Bioethanol B – Biobutanol.

Table 2.
Various proportions of diesel-ethanol blends by varying butanol from 0 to 10%.

These biobutanol-diesel blends were taken in a beaker for blending of bioethanol. Bioethanol was filled in burette and slowly added into biobutanol-diesel blends in the proportions ranging from 0 to 50% in increments of 5% of bioethanol assisted with magnetic stirring. This was carefully handled such that bioethanol will not evaporate during the process of blending. The magnetic stirrer (**Figure 1**) has been operated at a speed of 1500 rpm and for a set cycle of 2 minutes.

Each blend has been stirred for three to five times and the prepared blends were kept in a temperature controlled box (**Figure 2**) for five different temperatures 5, 10, 15, 20 and 25°C. This temperature range has been chosen by considering the climatic conditions of India. In India most part of the country [13] will attain 5°C during the winter season. The fuel blend found by this study has to be suitable to fuel CI engine for most places in our country. Fuel blends after the temperature stability tests are presented in **Figure 3**. **Figure 3** show three representative samples which are kept at 5°C (**Figure 3**) for a period of 20 days.

Periodical monitoring has been performed and the statuses of the blends were recorded. This is to find out the homogeneity of the fuel blend and to ensure that there is no phase separation between diesel and alcohols.

2.2 Property test

Prepared fuel blends (100 blends) were tested for the essential properties required as per the ASTM standards and comparison made [14] with respect to the diesel fuel as base. The instruments used for the properties along with the accuracy and ASTM standards are listed in Table. **Table 3** lists the properties of 5 representative fuel blends [15] containing 15, 25, .35, 45 and 50% of ethanol in comparison to that of diesel.

Figure 1.
Magnetic stirrer used for the blend preparation.

Figure 2.
Temperature control box for storing the prepared blend in various temperatures.

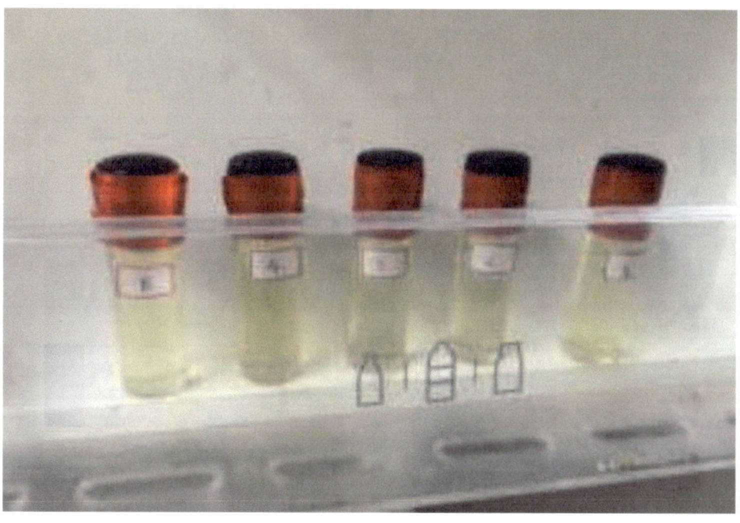

Figure 3.
Samples kept at 5°C after 20 days.

2.3 Experiment set up

A water-cooled, direct injection, Kirloskar make diesel engine [16] of 4.4 kW capacity at the rated speed 1500 rpm was used for testing the fuel blends. The engine (**Figure 4**) is coupled with eddy current dynamometer and electrical loading. Fuel flow was measured using burette and digital stop watch. Intake air flow was monitored by manometer and orifice plate. The displacement volume was 661.5CC, compression ratio was 17.5:1 and the recommended nozzle opening pressure was set at 200–205 bar. AVL pressure sensor has been used to capture the

S.No.	Property	Unit	Instrument Used	Accuracy	Percentage of Uncertainty	ASTM Standard
1.	Flash Point,	°C	Pensky-Martens Closed cup	± 0.1 °C	± 0.05%	ASTMD93-16a
2.	Kinematic Viscosity	mm^2/sec	Red wood viscometer	0.01 Centi Stokes	±0.02%	ASTMD445/446
3.	Calorific Value	kJ/kg	Bomb Calorimeter	1 J/grams.	±0.1%	ASTMD4868
3.	Cetane Number	No Unit	Ignition delay	± 0.1	±0.07%	ASTMD976/ ASTMD4737

Table 3.
List of instrument used for property testing.

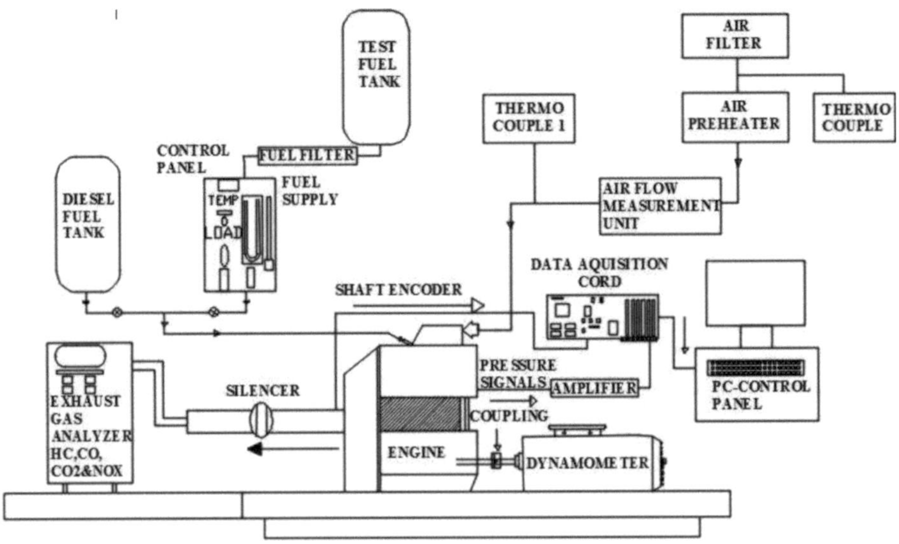

Figure 4.
Schematic layout of experimental set up.

pressure during the cylinder operation and to feed the captured signals to the data acquisition device. The fuel system of the experimental set up was mechanically controlled type and this was periodically cleaned and calibrated as per the recommendations of the manufacturers. Air preheater for heating the incoming air was used in the suction side. A heater of coil type of 1.0kVA capacity has been deployed heating the incoming air. The input electrical supply has been varied by a power regulator installed with the heater to obtain temperature difference. The input and out condition of the air has been measured by two separate thermocouple enabled with electronic readout. The injector used for fuel injection is a jet injector and is of mechanical type with a proper calibration. The average nozzle opening pressure has been set at 200 bar. To vary the nozzle opening pressure a washer of 0.20 mm has bee deployed in the vicinity of nozzle and the spring in the injector. The nozzle opening pressure has been verified with a dial type calibrating gauge for ensuring the pressure. The statndard injection timing of the injectore has been set as 23 deg. before top dead centre. To obtain the varied injection timing a washer has been added which is of 0.25 mm thickness. This was provided in vicinity of the engine and the fuel feed pump. This washer has been procured from the manufacturer and deployed as per the guidelines dictated by the supplier. Data acquisition system used for the present study consists of a computer, programmed with AVL 621 IndiModul system, which is receiving the signals amplified by a charge amplifier from a water cooled pressure transducer of KISTLER piezo electric transducer.

This system was controlled by IndiCom software. Specifications of the pressure transducer are given in Appendix 4. This device was programmed for generating the combustion data according to the pressure input. The encoder captures the position of the crank angle of the respective pressure signal and was duly connected to the engine. Specifications of the encoder are tabulated. 200 cycles of pressure data were captured and recorded for the analysis of combustion characteristics in the data acquisition system. This combustion parameters calculations were performed from the input received from the pressure transducer, crank angle encoder and intake air measurement. This also receives the input from the thermocouples for the temperature of the intake air, exhaust gases and incylinder. AVL-444 Di-Gas analyzer is used in this study for capturing the emissions from the test engine fuelled by the blends during the experiment. This measures CO, HC, NOx, and CO2 and oxygen concentration. It uses non-dispersive infrared (NDIR) sensor for measuring CO, $CO_2$and HC. Also, it measures NOx and oxygen concentrations by electrochemical sensors. All the emissions are recorded and converted to g/kWh for further analysis. This device is auto calibrated periodically as per the manufacturer advice. The measured values from the exhaust gas analyzer are in ppm [17] and the following conversion equations depict the onversion of ppm to g/kWhr which are standard equations (assuming 5% residual oxygen).

1000 ppm of NOx corresponds to 6.60 g/kWh.

100 ppm of HC corresponds to 0.20 g/kWh.

100 ppm of CO corresponds to 0.36 g/kWh.

2.4 Experimental uncertainty

Any experiment has its own uncertainty and the overall uncertainty has been arrived from individual uncertainties of the various instruments used. In the present study various instruments have been used and each one has different level of uncertainty. Hence, a detailed uncertainty analysis was carried out by the method of [18]. The total was arrived as ±1.3%. The uncertainty in any measured parameter was estimated based on Gaussian distribution method with confidence limits of $\pm 2\sigma$ (95% of measured data lie within the limits of 2σ of mean). Thus the uncertainty (Eq. (1)) was estimated using the following equation:

Uncertainty of any measured parameter

$$(\Delta x_i) = \left(2\sigma_i/\overline{X_i}\right) * 100 \tag{1}$$

From the uncertainties of the measured parameters, the uncertainties in computed parameters are evaluated by using an expression, which is derived as follows. If an estimated quantity, R depends on independent variable like (x_1, x_2, x_3 … … . x_n) then the error in the value of "R" is given by Eq. (2).

$$\Delta R = \left[\left(\frac{\partial R}{\partial x_1}\Delta x_1\right)^2 + \left(\frac{\partial R}{\partial x_2}\Delta x_2\right)^2 + \dots \dots \dots \dots \dots \dots + \left(\frac{\partial R}{\partial x_n}\Delta x_n\right)^2\right]^{1/2} \tag{2}$$

The estimated uncertainty values at a typical operating condition are given below:

Speed: ± 0.12% Load: ± 0.49%
Mass flow rate of air: ± 0.62% Mass flow rate of diesel: ± 0.87%
Brake power: ± 0.25% Brake thermal efficiency:± 0.27%
NO_X: ± 1.1% Hydrocarbon: ± 0.01%
CO: ± 0.8% Smoke: ± 1.3%

There are various methods available to reduce the errors observed in the instruments such as selecting the instruments according to the measurement level required (range of measurement), Accuracy of the instrument used, sensitivity etc., this experiment was conducted by deploying the appropriate instruments within the range of measurement, accuracy and sensitivity requirement.

3. Results and discussion

From **Figure 5** it is seen that the two blends D75E15B10 and D65E25B10 offers higher incylinder pressure in comparison to that of diesel. This can be attributed to the improvement in the physicochemical properties of the blends up to a certain extent of ethanol into diesel. This is due to the improved complete combustion of the blends by the addition of ethanol till 25% by volume (**Figure 6**).

However, fuel blends D55E35B10 and D45E45B10 produce lesser in cylinder pressure compared to diesel [18]. This is due to the suppression of combustion by the higher volume of ethanol in the blends, which is due to higher heat of vaporization. It can also be observed from figure that the peak pressure from D75E15B10 and D65E25B10 are 6.4% and 15.2% higher than diesel. **Figure** 7 shows the variation of incylinder peak pressure versus brake power for the blends. It is seen that the addition of ethanol (up to a volume of 25%) into diesel increases the incylinder peak pressure. Significantly. Also, the increase in the incylinder peak pressure is found proportional to the increase in brake power. This is due to the improvement in the physico-chemical properties of the blends by the addition of ethanol. The improvement in kinematic viscosity [19], density results in better atomization which leads

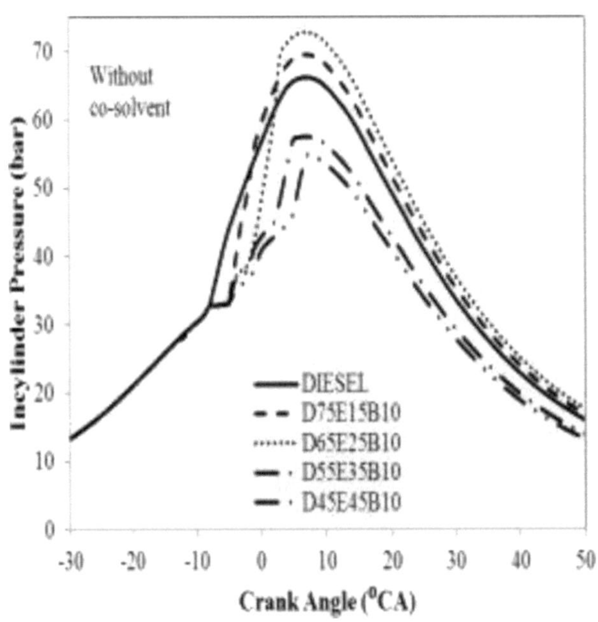

Figure 5.
Variation of Incylinder pressure versus crank angle.

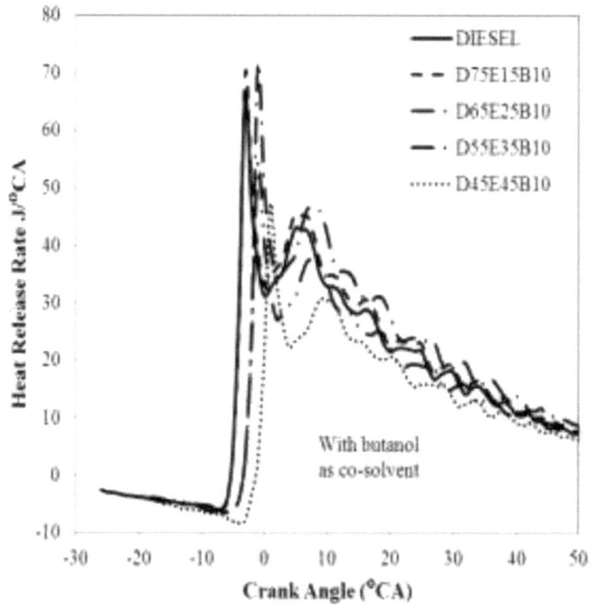

Figure 6.
Variation of HRR of fuel blends with crank angle at rated power.

to the more complete combustion. It is also seen that the addition of ethanol into diesel higher than 25% by vol. reduces the incylinder peak pressure significantly. This is due to the dominance of heat of vaporization of the blends with the increase in the volume of ethanol in the blend. This produces a cooling effect which results in poor atomization and lesser rate of oxidation which results in lesser incylinder peak pressure. The increases in incylinder peak pressure of D75E15B10 and D65E25B10 significantly. Also, the increase in the incylinder peak pressure is found proportional to the increase in brake power. This is due to the improvement in the physico-chemical properties of the blends by the addition of ethanol.

Heat release rate is an indicator of combustion efficiency and these parameters is helping for explaining the BTE, EGT, increase in the incylinder pressure, emissions from the engine and the pressure during the engine operation. **Figure 8** depicts the heat release rate diagrams generated during the engine operation fueled with the blends with and without butanol. The representation of a diagram generated at rated power condition is presented. HRR graphs are generated at all loads and for representation HRR at full load condition is presented. It can be seen from figure that ethanol addition up to 25% increases the HRR to a greater extent due to the enhanced combustion behavior resulted from better atomization. However, the increase in ethanol content beyond 25% decreases HRR of the blends as poor atomization resulting in lesser heat release rate (**Figure 9**). It is also seen that the volume of ethanol are directly having impact in this parameter. Increase in the volume of ethanol decreases the heat release rate. Longer crank angle has been necessary for the blend to start up the heat release rate compared to diesel. Also, these blends produced low combustion duration compared to diesel [20, 21]. It can be observed from the **Figure 10** that two blends D75E15B10 (22 Wang *et al.* 2018) offers a significant increase in BTE with respect to diesel under all brake power conditions. This can be attributed to the increase rate of spray characteristics which is a result of increase in the volatility of the fuel. This is the main reason for the increase in BTE. Beyond 25% of ethanol a decrease in heat release rate has been observed (**Figure 11**). This is due to the decrease in the self-ignition property which

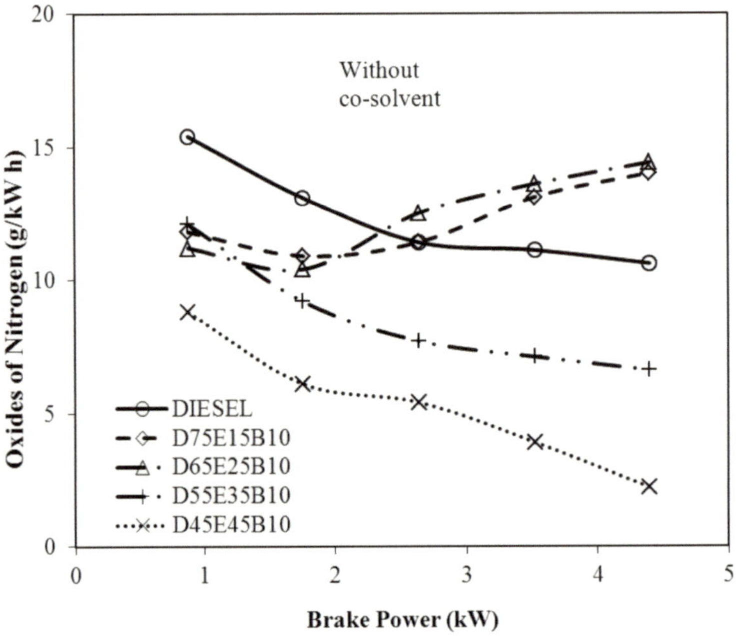

Figure 7.
Variation of NOx emissions of fuel blends versus brake power.

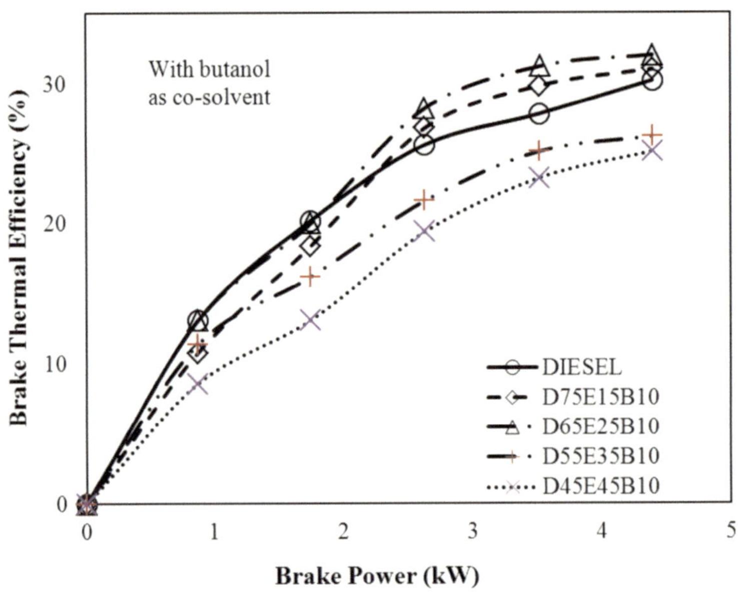

Figure 8.
Variation of brake thermal efficiency versus brake power.

is suppressed by the higher volume of ethanol. Variations of EGT versus brake power for the blends are as shown in **Figure 12**. It is seen that the EGT of the blends containing combustion characteristics by the dominance of heat of vaporization of the final blends. This produces a cooling effect in the incylinder which reduces the rate of reaction of fuel particles with oxygen available and results in lesser BTE and lesser EGT. This is the main reason for the reduction of EGT of D55E35B10 and D45E45B10 [22]. The increases of EGT for the blends D75E15B10 and D65E25B10

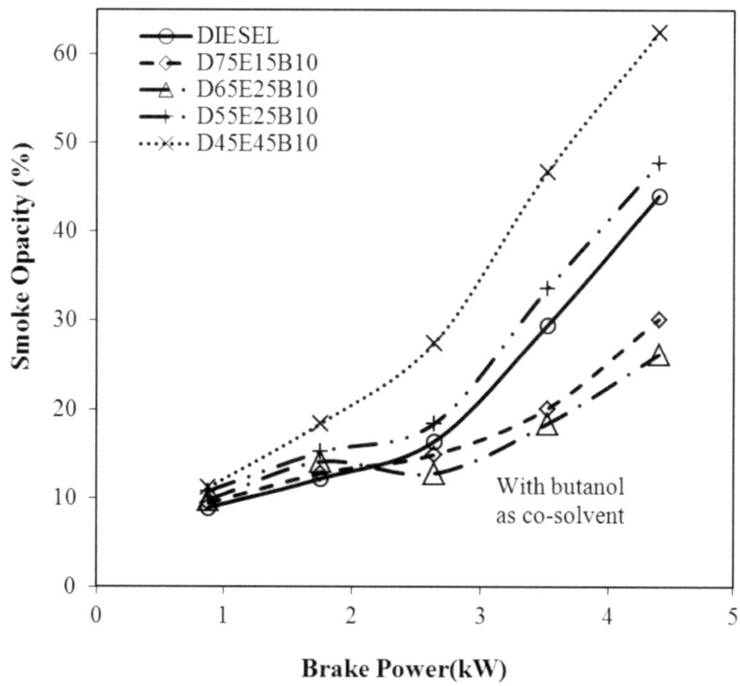

Figure 9.
Variation of smoke emissions versus brake power.

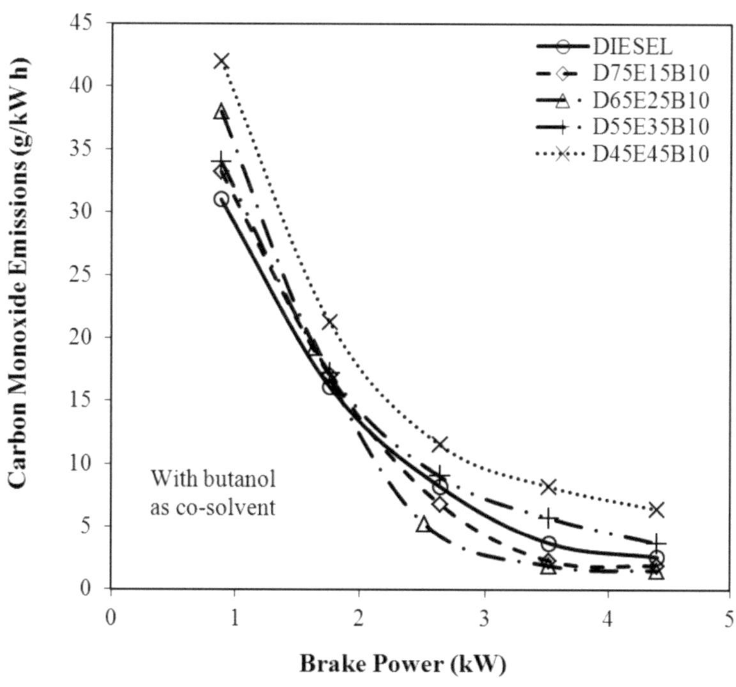

Figure 10.
Variation of CO emissions versus brake power.

are 16.9% and 22.6% respectively in comparison to diesel at rated power. The decrease of EGT for the blends D55E35B10 and D45E45B10 are 13.6% and 20.4% respectively in comparison to diesel at rated power. From **Figure 13** it is observed

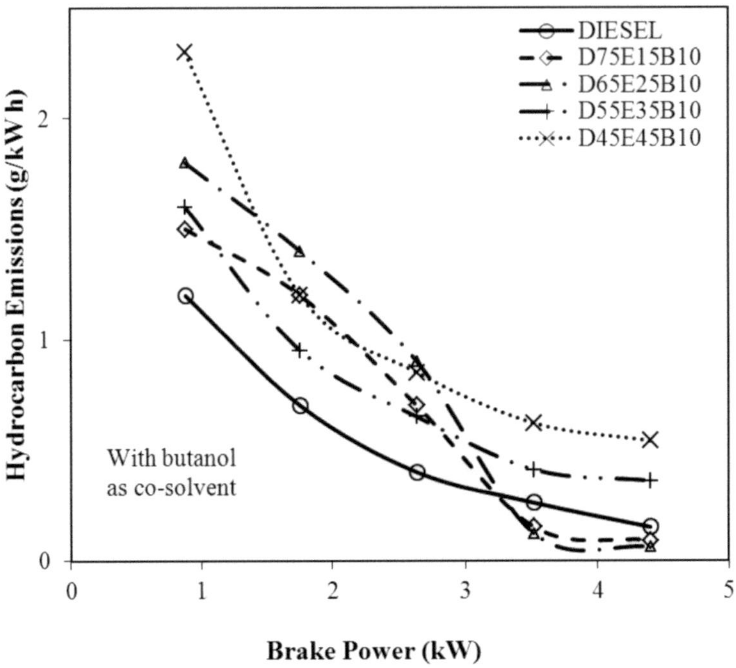

Figure 11.
Variations of HC emissions of fuel blends.

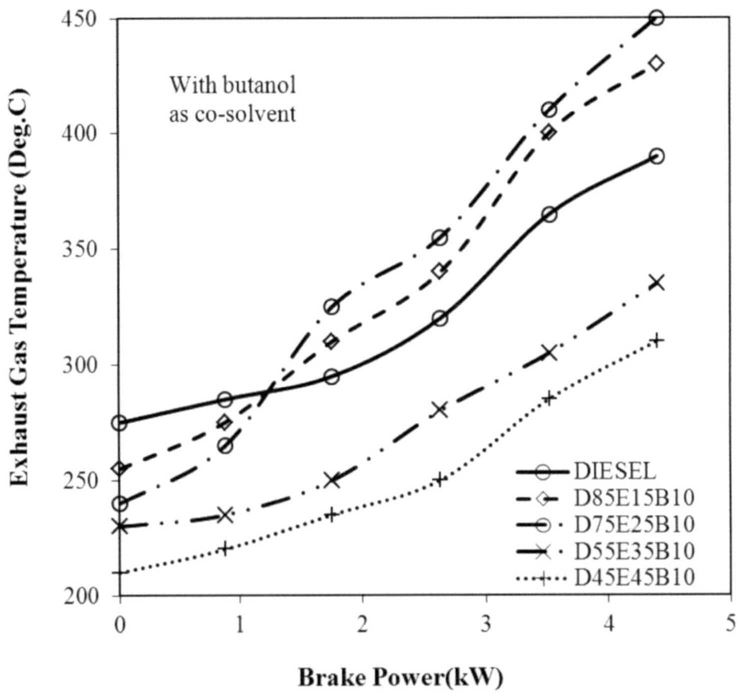

Figure 12.
Variations of exhaust gas temperature versus brake power.

that the blends containing lower ethanol content (lower than 25%) producing higher oxides of nitrogen and higher volume of ethanol content (higher than 25%) produce low oxides of nitrogen with respect to diesel. The ethanol addition

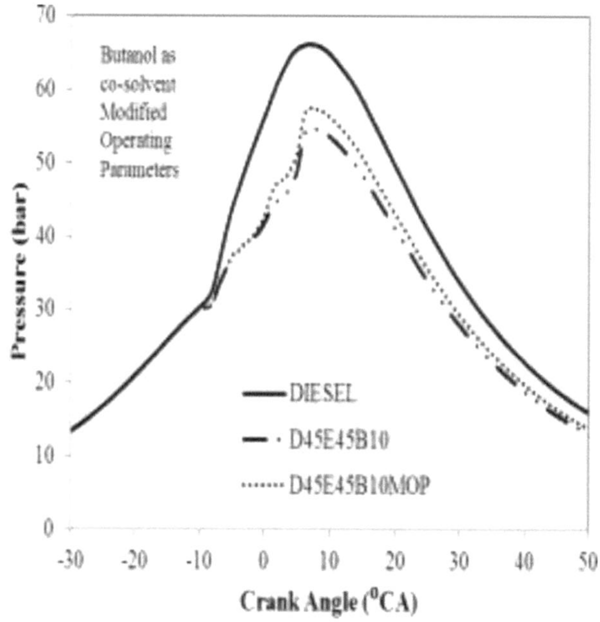

Figure 13.
Variation of cylinder pressure with crank angle at rated power.

improves the physico chemical properties such ease of evaporation, rate of atomization and a decrease in viscosity [23].

Figure 13 depicts the emissions of smoke for the engine fueled with various proportions of ethanol. Out of the fule blends the blend containing 25% ethanol produces the lowest smoke emission. This can be attributed to the enhanced physicochemical properties of the fuel blend up to this volume of ethanol. This increases the rate of combustion and resulted in a significant decrease in smoke emissions. Beyond this volume the dominance of the heat of vaporization suppresses the incylinder temperature and results in a decrease of oxidation rate and thereby higher smoke emissions. The present work utilizes various proportions of ethanol into diesel as fuel. **Figure 14** indicates that higher volume of ethanol produces higher smoke emissions and lower volume of ethanol (lower than 25%) produces lower smoke emissions. This can be attributed to increase in the rate of combustion up to 25% of ethanol and suppression of temperature above 25% of ethanol content in the blend The decreases of CO emissions from D75E15B10 and D65E25B10 are found as 27% and 46.1% respectively at rated power in comparison with diesel. The increases CO emissions from D55E35B10 and D45E45B10 are found as 37.2% and 58.2% respectively at rated power in comparison with diesel. From the **Figure 15** it can be seen that containing 15% of ethanol offers lower HC emissions compared to diesel. This is mainly due to the increase in combustion characteristics of the blends containing lower volume of ethanol and decrease in combustion characteristics of the blends containing higher volume of ethanol. From the **Figure 15**, it can be observed that D75E15B10 and D65E25B10 produce 28% and 7.6% lesser than diesel. The blend containing 45% of ethanol offers higher HC emissions and the blends. However, D55E35B10 and D45E45B10 produce significantly higher emissions compared to diesel at load conditions. However, D55E35B10 and D45E45B10 offer 8.2% and 12.6% lesser in cylinder pressure compared to diesel at full load. Also it can be observed that start of pressure rise of all fuel blends are away from that of diesel. This is due to lesser cetane number of the final blend compared to diesel. The previous phase of the present study indicated that D45E45B10 blend is containing

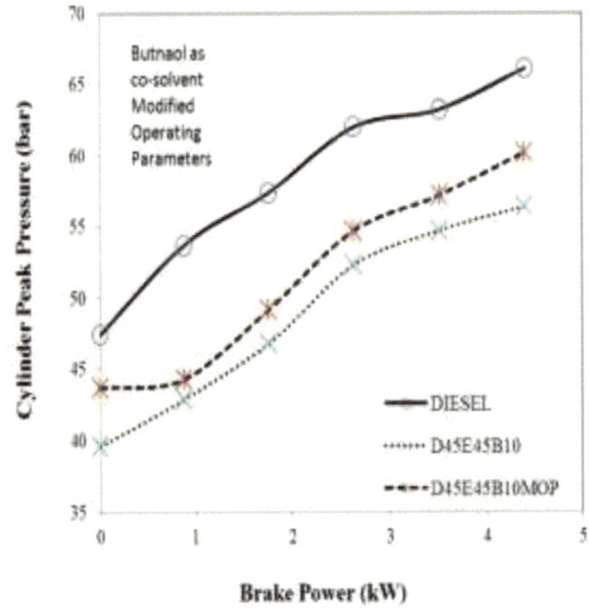

Figure 14.
Variation of Incylinder peak pressure versus brake power.

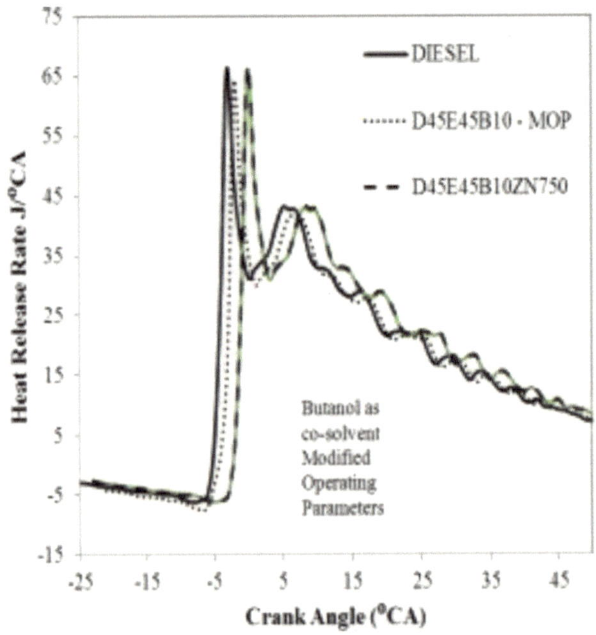

Figure 15.
Variation of HRR with crank angle at rated power.

higher possible volume of ethanol has failed to produce better performance and emission characteristics. Hence, it is decided to modify the parameters of the test engine to improve the performance and emission characteristics of the fuel blend. Also in the first phase this blend has not suffered phase separation which is the major limitation of utilizing ethanol diesel blends in CI engine up to a temperature of 5°C. The suitable parameters for fuelling CI engine by D45E45B10 have been determined by Taguchi method on ANOM approach (Analysis of mean) (**Table 4**).

Blend	Flash Point	Energy Content	Density	Kinematic Viscosity	Oxygen content	Cetane Number
Units	**^{0}C**	**MJ/l**	**kg/ m^3**	**mm^2 /s**	**wt..%**	
Diesel	74	42.8	829	4.04	0	50
Ethanol	13	26.9	790	1.37	34.8	8
n-butanol	35	33.1	809	3.2	21.58	25
D75E15B10	64	40.24	823	3.7	5.64	43.3
D65E25B10	57.9	38.65	818	3.45	9.12	39.1
D55E35B10	51.8	37.06	813	3.19	12.59	34.9
D45E45B10	47.5	37.13	807	2.94	17.16	30.7
D40E50B10	39.6	33.88	805	2.62	19.6	26.5

Table 4.
Properties of diesel-ethanol-butanol blends.

This part of the work used Taguchi method for designing experimental layout and rank matrix to attain optimum level of parameters.

The steps involved in the optimization process are:

- Selection of operating parameters and their levels
- Selection of Orthogonal array by Taguchi method
- Preparation of experimental layout
- Conducting the experiments using the experimental layout
- Observation of response parameters
- Listing the results and formation of Rank matrix
- Suggesting optimal level of parameters
- Conducting engine experiment using optimal parameters

Present investigation has considered four operating parameters viz. injection pressure (IP), injection timing (IT), compression ratio (CR) and intake air temperature (IAT) for optimization. The range and level of parameters are decided with literature support and preliminary engine experiments. **Table 5** shows the level of operating parameters.

S.No.	Symbol	Parameters	Level 1	Level 2	Level 3
1	A	Injection Pressure (IP) (bar)	190	200	210
2	B	Injection Timing (IT) (^{0}BTDC)	26	29	32
3	C	Compression Ratio (CR)	17.5	19	21
4	D	Intake Air Temperature(IAT) (°C)	50	75	100

Table 5.
Parameters involved in the optimization and their levels.

Using this parameters and their levels a suitable orthogonal array, experimental layout and number trials of the experiments have been arrived from Taguchi method of optimization.

Taguchi method of optimization offers a systematic approach to arrive at the level of performance parameters involved in the response parameters. The Taguchi method uses an orthogonal array for designing the experimental layout. The selection of orthogonal array is arrived from the degrees of freedom of the parameters involved. The minimum number of experiments (trials) for selecting the optimum level of parameters can be determined using the relation:

$$N = (L - 1) * P + 1 \quad (3)$$

Where, N = Total number of test runs, L = Number of levels of parameters and P = Number of control parameters.

The present study uses (**Table 5**) four parameters and three levels and hence, the total degrees of freedom of control parameters are 8. Therefore, L_9 is suitable OA for the total degrees of freedom of involved parameters.

Analysis of Mean (ANOM) This is used after attaining the experimental results as per the L_9 orthogonal array of nine experiments containing 3 sets of reading in each setting. A rank matrix table is utilized for the analysis of captured data (**Table 6**). A rank matrix **Table 7** has been constructed to arrive at the optimal level of parameters. Average of the sum of the each level outcome has been obtained and the rank is tabulated for the maximum of the outcome. Assuming that Y as output parameter and the level summation has been obtained as:

Trial No.		**Column No.**		
	A	**B**	**C**	**D**
1.	1	1	1	1
2.	1	2	2	2
3.	1	3	3	3
4.	2	1	2	3
5.	2	2	3	1
6.	2	3	1	2
7.	3	1	3	2
8.	3	2	1	3
9.	3	3	2	1

Table 6.
L_9 orthogonal array.

Rank	**A_1 (Level 1)**	**B_2 (Level 3)**	**C_3 (Level 3)**	**D_4 (Level 3)**
Level/Parameter	**A**	**B**	**C**	**D**
1	31.3	31.6	30.6	30
2	31.05	29.86	31.36	31.6
3	31.1	31.96	31.66	31.7

Table 7.
Rank Matrix (for BTE).

$$A_1 = Y_1 + Y_2 + Y_3 \text{ (in which the level 1 is denoted in the orthogonal array)} \quad (4)$$

Similar calculation has been done for three levels and for four parameters, from which the rank matrix table has been constructed:

From the **Table 7** it can be concluded that IP 190 bar (LEVEL1), IT 29^0bTDC (LEVEL3), CR 19 (LEVEL3) and IAT 100 (LEVEL3) are the optimal parameters by comparing the rank. The same sets of readings are captured for NOx to match with the brake thermal efficiency. The optimized levels of operating parameters are as shown in **Table 7**. Blend D45E45B10 have been tested under the modified operating parameters and the results are compared with diesel and D45E45B10 under normal operating parameters. The same engine has been used for the testing of the blends under modified operating parameters. The results of the experiment are presented in graphical form. The variation of cylinder pressure with crank angle at rated power for the blend D45E45B10 under standard operating parameters and modified operating parameters are presented in **Figure 13**, it is seen that the modified engine operating parameters increased the cylinder pressure significantly compared to diesel. This is due to the increased heat energy release in the combustion chamber with increase in compression ratio and intake air temperature. Also, the advancement in the injection timing improves the precombustion phase and results in more complete combustion. This shows the suitability of the modified engine operating parameters for the blend D45E45B10. The increase in pressure of D45E45B10MOP is found as 7.1% higher than diesel at rated power. However, the cylinder pressure is found lesser than diesel. This is due to the lesser essential properties of D45E45B10 in comparison to diesel. Variation of incylinder peak pressure versus brake power for D45E45B10 under modified engine operating parameters is shown in **Figure 14**. It is seen that the incylinder peak pressure increases by fuelling D45E45B10 under modified operating parameters compared to that of normal operating parameters. This is due to the suitability of the modified operating parameters for the blend D45E45B10. Also, the increase in the incylinder peak pressure is found proportional to the increase in brake power. This increase is due to the improved rate of combustion by the increase in compression ratio and intake air temperature. Also, the advancement of injection timing improved the precombustion phase which suppresses the dominance of heat of vaporization of the blend. However, the incylinder peak pressure of D45E45B10MOP is found lesser than diesel at all load conditions. This is due to the lesser energy content of D45E45B10 in comparison to diesel. The increase in the incylinder peak pressure of D45E45B10MOP is found as 6.3% higher than D45E45B10. Heat release rate is an indicator of combustion efficiency and these parameters is helping for explaining the BTE, exhaust gas temperature, rate of pressure rise, emission parameters and cylinder pressure.

Figure 15 shows the It can be seen from figure that ethanol addition up to 25% increases the HRR to a greater extent due to the enhanced combustion behavior resulted from better atomization. However, the increase in ethanol content beyond 25% decreases HRR of the blends as poor atomization resulting in lesser heat release rate.

From the **Figure 16** it is observed that the target blend D45E45B10 offers higher BTE with modified operating parameters compared to that of BTE with normal operating parameters. However, this blend offers lesser BTE compared to that of diesel (**Figure 17**). The reason for the increase in BTE is due to the increase in heat content of the combustion chamber resulted from the enhanced combustion triggered by the modified operating parameters. Ignition quality, which decreases the combustion temperature and thereby lesser BTE compared to diesel. Similar

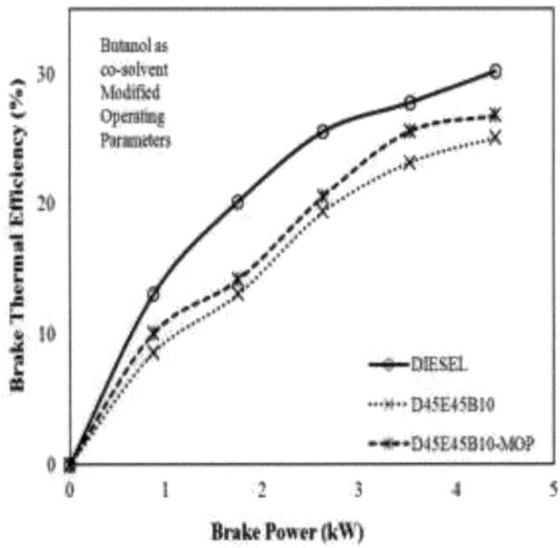

Figure 16.
Variation of brake thermal efficiency versus brake power.

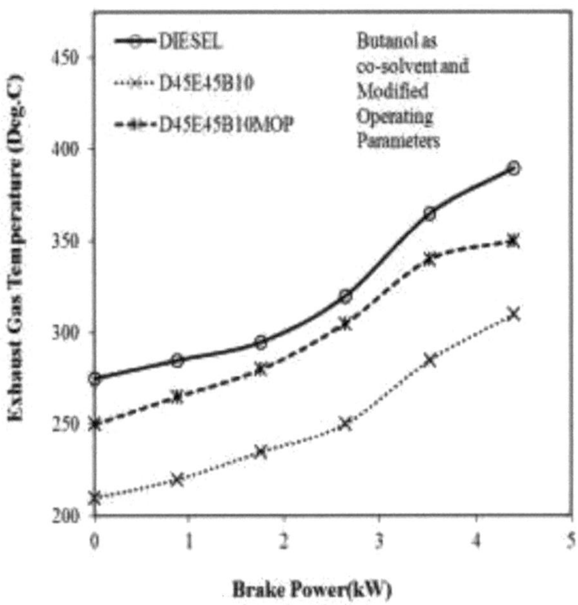

Figure 17.
Variation of EGT versus brake power.

observation was presented by previous researchers [24]. The increase in BTE by the modification of operating parameters is 6.7% compared to those in normal operating parameters, which indicates the suitability of the parameters for the target blend. The decrease in BTE of the target blend at modified operating parameters is only 2.1% compared to diesel. Variation of EGT with respect to brake power is as shown in **Figure 18**. The quantity of ethanol in the blend determines the performance of the blend as the increase in ethanol volume results in poor to brake power for the blend D45E45B10 operated under normal operating parameters and modified operating parameters in comparison to diesel. It is seen that there is a

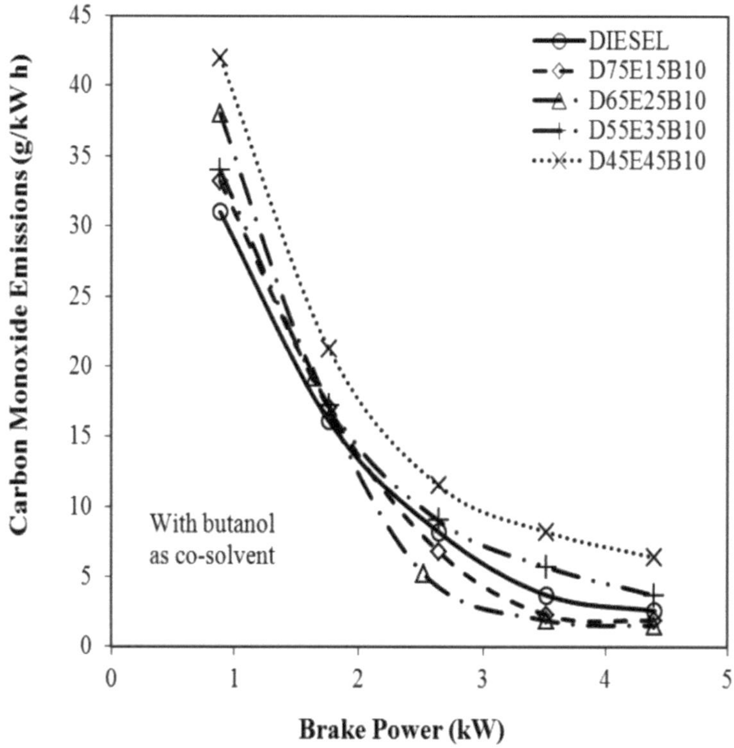

Figure 18.
Variations of CO emissions versus brake power.

significant increase in EGT of D45E45B10MOP in all load conditions compared those under normal operating parameters. This is due to the higher heat energy release by the blend operated under modified operating parameters. This is due to the suppression of the dominance created by the heat of vaporization of the higher volume of ethanol by the modified parameters to a certain extent. However, the EGT of D45E45B10MOP is found lesser than diesel. The increase of EGT of D45E45B10MOP is found 13.1% higher than D45E45B10 at rated power.

Any engine producing higher emissions of oxides of nitrogen is an indication of higher temperature of the in cylinder which is the result of complete combustion. **Figure 19** shows the emissions of oxides of nitrogen from the engine fueled using D45E45B10 with and without modification of parameters along with diesel for comparison. It can be observed that there is an increase in NOx emissions from the target blend when fuelled in CI engine which shows the suitability of the modified parameters. This is due to the increase in heat content of the target blend operating with modified operating parameter and compressed air, which helps to combust the fuel by reducing the ignition delay. However, the emissions of NOx are lesser than diesel as the higher volume of ethanol suppresses the temperature of the in cylinder. The increase in NOx emissions due to the modification of operating parameters is 100% (approximately double) compared to that of operating under normal operating parameters. The decrease in NOx emissions of D45E45B10 –MOP is 40.5% compared to that of diesel at full load condition. **Figure 20** shows the smoke opacity of the target blend under modified operating parameters at all load conditions. It can be observed that there is a significant reduction in smoke emissions from the target blend under modified operating parameters compared to that under normal operating parameters. This is due to the reason of increased temperature of the in

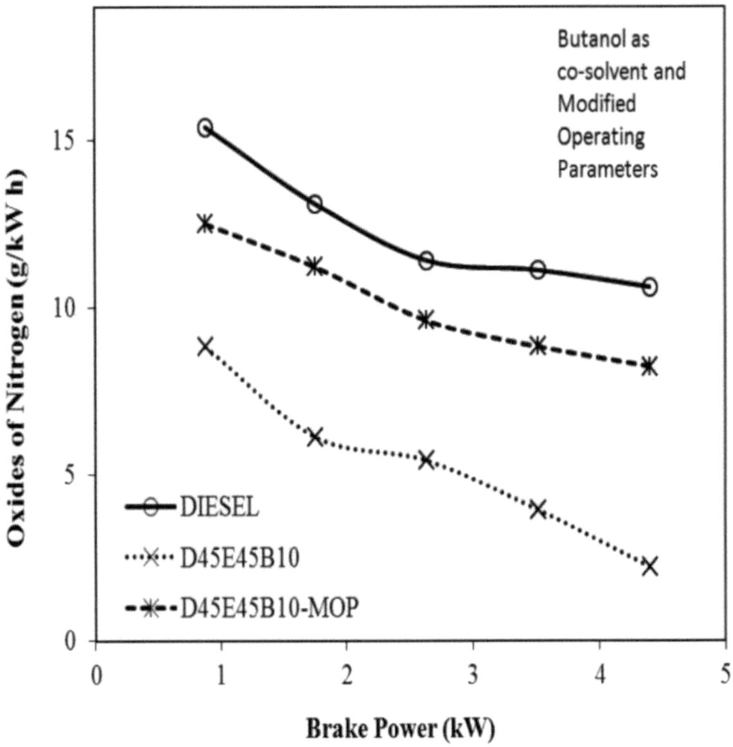

Figure 19.
Variation of NOx emissions versus brake power.

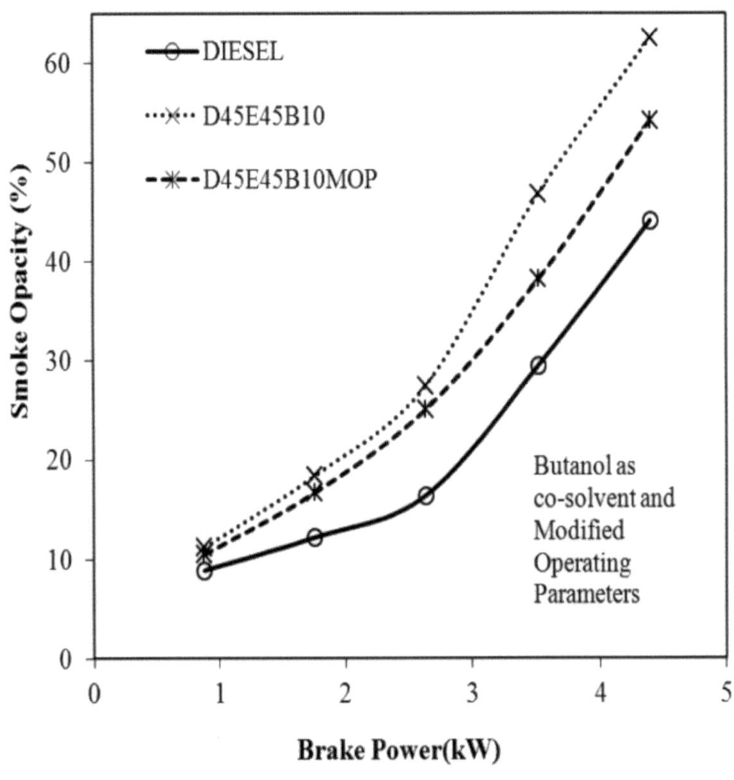

Figure 20.
Variation of smoke opacity versus brake power.

cylinder by the modified operating parameters which enhances higher heat release resulted from compressed air. However, the higher heat of vaporization of the blend still suppresses the temperature and hence there is an increase in smoke emissions compared to that of diesel. The decrease in smoke emissions is 21.2% compared to D45E45B10 operated under normal operating parameters. The increase in smoke emissions of D45E45B10-MOP is 16.5% higher than diesel at full load condition. Similar results were observed by previous researchers [25]. From the **Figure 18** it can be seen that there is a significant reduction of CO emissions due to the modification of operating parameters to the target blend. This is due to impact of the modified parameters on the combustion characteristics to a certain extent. However, the higher ethanol content increases the heat of vaporization of the final blend, which results in poor ignition quality which results in lesser temperature of the in cylinder shows the variation of CO emissions of D45E45B10 fuelled in the test engine under modified operating parameters compared to that of diesel. This reduces the BTE of the blend lesser than diesel.

The increase in BTE of the blend at modified operating parameters is 29.6% compared to that operated under normal operating parameters. However, the increase in CO of the blend is 19.3% higher than diesel. Higher ethanol content affects the self-ignition property; hence it reduces reaction rate, combustion temperature and heat release rate [26].

4. Conclusion

Different phases of study have been followed to utilize diesel ethanol blends as fuel in compression ignition (CI) engine in this study. Experiments were conducted with diesel ethanol without co- solvent and with butanol as co-solvent. The effects of engine operating parameters such as Injection Pressure (IP), Injection Timing (IT), Compression Ratio (CR) and Intake Air Temperature (IAT) on engine performance, combustion and emission were studied.

- Solubility test indicates that ethanol can be blended with diesel up to a volume of 50% with 10% butanol as co-solvent. This blend is found as stable up to a lower temperature of 5°C for 20 days.

- Property testing show that properties of the blend containing 45%of ethanol and 10% butanol as co-solvent is found suitable for replacing diesel to fuel CI engine.

- However, blend containing 50% ethanol and 10% butanol is found not suitable as the cetane number is less than 30 which is a minimum requirement as per ASTM standards.

- Improved physicochemical properties, Better ignition quality, higher combustion temperature and higher oxygen content increase the NO_x emission by 13.2% in the case of D80E20 whereas 2.9% increase is observed for D90E10 compared with diesel.

- The smoke level of D80E20 49.2% lesser than that of diesel but D90E10 results in 32% lesser smoke emission.

- The HC and CO emissions are reduced by about 66% and 9.6% respectively in D80E20 operation compared to diesel. Improved atomization of fuel in the incylinder in lesser HC and CO emission than diesel.

- The lower cetane number of D45E45B10 retard the combustion by 4°CA compared to diesel operation.
- The peak pressure is lower for D45E45B10 in the entire load range when compared to diesel operation.
- The lower energy content and higher heat of vaporization of D45E45B10 leads a lower peak heat release rate compared to diesel fuel operation.
- The D45E45B10 shows a significantly lower brake thermal efficiency compared to diesel operation and is found 16.8% lesser than diesel at rated power.
- The NO_x emission for D45E45B10 is 22.5% lesser than diesel operation due to lower energy content and higher heat of vaporization.
- The increase in smoke emission is about 49.2% for D45E45B10 operation compared to diesel.
- HC is increased by 6.7% in the case of D45E45B10 operation compared to diesel operation. The CO emission follows the same trend as that of HC emission.
- Even though this phase gave adverse effects in performance and emissions, higher volume of ethanol is utilized without any phase separation.
- D45E45B10 can be used as a fuel for CI engine with modified operating parameters. This enhanced improved complete combustion and shows significant improvement in performance at all load conditions.
- D45E45B10MOP operation advances the combustion and improves premixed combustion compared to D45E45B10 under normal operating parameters. However, D45E45B10 shows lower peak heat release rate and peak pressure at rated power compared to diesel operation.
- D45E45B10MOP operation improves the BTE significantly compared to D45E45B10 at all load conditions. The increase in BTE at rated power is 6.8% higher than D45E45B10 fuelled under normal operating parameters. However, BTE of D45E45B10MOP is found lesser than diesel at rated power.
- There is a significant increase in NO_x emission in D45E45B10MOP operation compared to D45E45B10. The increase in NOx emissions of D45E45B10MOP is found thrice that of NOx emissions from D45E45B10 fuelled under normal operating parameters. However, NOx emissions of D45E45B10MOP are found lesser than diesel at rated power.
- The smoke emission is reduced by 15.4% in fueling D45E45B10MOP compared to D45E45B10 fuelled under normal operating parameters. However, the smoke emissions from D45E45B10MOP are found higher than diesel at rated power.
- The HC and CO emissions are reduced by 22.5% and 9.2% respectively in fuelling D45E45B10MOP compared to D45E45B10 fuelled under normal

operating parameters. However, this is found higher than diesel at all load conditions which is due to physicochemical properties of D45E45B10 compared to diesel.

As a sum up, although the efficiency produced by D45E45B10 is found to be marginally lower and the emissions of smoke, HC & CO produced are found to be marginally higher compared to that of diesel. The utilized ethanol and butanol are manufactured from waste products and the emissions of oxides of nitrogen produced are found to be significantly lower compare to that of diesel. Hence, higher volume of ethanol can be utilized and a saving of 55% of diesel fuel can be achieved by the implementation of this modification in fuel and in engine. This in turn reduces the dependency of other countries for import of crude oil.

Author details

Prabakaran Balasubramanian
Department of Automobile Engineering, Hindustan Institute of Technology and Science, Chennai Tamil Nadu India

*Address all correspondence to: b7prabakaran@gmail.com

References

[1] Han, J., Somers, L. M. T., Cracknell, R., Joedicke, A., Wardle, R., & Mohan, V. R. R. (2020). Experimental investigation of ethanol/diesel dual-fuel combustion in a heavy-duty diesel engine. Fuel, *275*, 117867.

[2] Mamat, R., Sani, M. S. M., Sudhakar, K., Kadarohman, A., & Sardjono, R. E. (2019). An overview of Higher alcohol and biodiesel as alternative fuels in engines. Energy Reports, *5*, 467–479.

[3] Kumar, S., Cho, J. H., Park, J., & Moon, I. (2013). Advances in diesel–alcohol blends and their effects on the performance and emissions of diesel engines. Renewable and Sustainable Energy Reviews, *22*, 46–72.

[4] Gao, Z., Lin, S., Ji, J., & Li, M. (2019). An experimental study on combustion performance and flame spread characteristics over liquid diesel and ethanol-diesel blended fuel. Energy, *170*, 349–355.

[5] Singh, G. N., & Bharj, R. S. (2019). Study of physical-chemical properties for 2nd generation ethanol-blended diesel fuel in India. Sustainable Chemistry and Pharmacy, *12*, 100130.

[6] Rakopoulos, C. D., Rakopoulos, D. C., Kosmadakis, G. M., & Papagiannakis, R. G. (2019). Experimental comparative assessment of butanol or ethanol diesel-fuel extenders impact on combustion features, cyclic irregularity, and regulated emissions balance in heavy-duty diesel engine. Energy, *174*, 1145–1157.

[7] Yu, Y. (2019). Experimental study on effects of ethanol-diesel fuel blended on spray characteristics under ultra-high injection pressure up to 350 MPa. Energy, *186*, 115768.

[8] Woo, C., Kook, S. and Hawkes, E.R. (2016).Effect of intake air temperature and common-rail pressure on ethanol combustion in a single-cylinder light-duty diesel engine. Fuel., 180: 9–19.

[9] Verma, S., Sharma, B., Ahmad, J., Dwivedi, G. and Nandan, G., 2018. Impact assessment of Ethanol as Fuel for Engine operation. Materials Today: Proceedings, *5*(2), pp.6115–6120

[10] Ribeiro, N.M., Pinto, A.C., Quintella, C.M., da Rocha, G.O., Teixeira, L.S., Guarieiro, L.L., do Carmo Rangel, M., Veloso, M.C., Rezende, M.J., Serpa da Cruz, R. and de Oliveira, A.M. (2007). The role of additives for diesel and diesel blended (ethanol or biodiesel) fuels: a review. Energy & fuels., 21(4): 2433–2445.

[11] Hafid, H. S., Shah, U. K. M., Baharuddin, A. S., & Ariff, A. B. (2017). Feasibility of using kitchen waste as future substrate for bioethanol production: a review. Renewable and Sustainable Energy Reviews, *74*, 671–686.

[12] Stoeberl, M., Werkmeister, R., Faulstich, M., & Russ, W. (2011). Biobutanol from food wastes–fermentative production, use as biofuel an the influence on the emissions. Procedia Food Science, *1*, 1867–1874.

[13] Hansen, A. C., Zhang, Q., & Lyne, P. W. (2005). Ethanol–diesel fuel blends—a review. Bioresource technology, *96*(3), 277–285.

[14] B.Prabakaran, P.,Vijayabalan. (2016).Evaluation of the Performance of n-butanol-ethanol-diesel blends in a diesel engine, International Journal of Energy for a Clean Environment **17:** 81–97.

[15] Prabakaran, B., Vijayabalan, P. and Balachandar, M. (2017). Experimental Investigation of Ethanol-Diesel-Butanol Blends in a Compression Ignition

Engine by Modifying the Operating Parameters. *SAE International Journal of Engines*, *11*(03–11-05-0037).

[16] Prabakaran, B., Vijayabalan, P., & Balachandar, M. (2019). An assessment of diesel ethanol blend fueled diesel engine characteristics using butanol as cosolvent for optimum operating parameters. *ENERGY SOURCES PART A-RECOVERY UTILIZATION AND ENVIRONMENTAL EFFECTS*.

[17] Holman, J.P. and Gajda, W.J., 2001. *Experimental methods for engineers* (Vol. 2). New York: McGraw-Hill.

[18] Gnanamoorthi, V. and Devaradjane, G. (2015). Effect of compression ratio on the performance, combustion and emission of DI diesel engine fueled with ethanol–diesel blend. Journal of the Energy Institute.,88(1): pp.19–26.

[19] Kuszewski, H. (2018). Experimental investigation of the effect of ambient gas temperature on the autoignition properties of ethanol–diesel fuel blends. Fuel, 214, pp.26–38.

[20] Pinzi, S., López, I., Leiva-Candia, D. E., Redel-Macías, M.D., Herreros, J.M., Cubero-Atienza, A. and Dorado, M.P. (2018). Castor oil enhanced effect on fuel ethanol-diesel fuel blend properties. Applied Energy, 224, pp.409–416.

[21] Yilmaz, N. (2012). Comparative analysis of biodiesel–ethanol–diesel and biodiesel–methanol–diesel blends in a diesel engine. Energy, 40(1), pp.210–213

[22] Verma, S., Sharma, B., Ahmad, J., Dwivedi, G. and Nandan, G., 2018. Impact assessment of Ethanol as Fuel for Engine operation. Materials Today: Proceedings, 5(2), pp.6115–6120

[23] Ghadikolaei, M.A., Cheung, C.S. and Yung, K.F., 2018. Study of combustion, performance and emissions of diesel engine fueled with diesel/biodiesel/alcohol blends having the same oxygen concentration. Energy

[24] Rakopoulos, D.C., Rakopoulos, C. D., Giakoumis, E.G., Papagiannakis, R. G. and Kyritsis, D.C. (2014). Influence of properties of various common bio-fuels on the combustion and emission characteristics of high-speed DI (direct injection) diesel engine: Vegetable oil, bio-diesel, ethanol, n-butanol, diethyl ether. Energy.,73: 354–366.

[25] D.C., Rakopoulos, Antonopoulos, K. A., Rakopoulos, D.C. and Hountalas, D. T. (2008). Multi-zone modeling of combustion and emissions formation in DI diesel engine operating on ethanol–diesel fuel blends. Energy conversion and management.,49(4): 625–643.

[26] Sayin, C. and Gumus, M. (2011). Impact of compression ratio and injection parameters on the performance and emissions of a DI diesel engine fueled with biodiesel-blended diesel fuel. Applied thermal engineering.,31(16): .3182–3188.

Chapter 6

Bioethanol-Diesel Blends Used in Diesel Engines and Vehicles under Transient Operation

Octavio Armas Vergel, Dolores Cárdenas, Reyes García-Contreras and Carmen Mata

Abstract

This book chapter describes, the experiences of different experimental works related to the study of the effect of bioethanol-diesel fuel blends directly used on a study about the integrity of some parts of a diesel injection system, and on performance and regulated emissions of engines and/or vehicles under different transient conditions. The experiences described are carried out in an engine test bench, two public busses under urban transportation and a construction machine in an actual railway construction. A bioethanol-diesel fuel blend with 10% v/v of bioethanol was used for engine testing to study the potential effect on performance and emissions. Later, a blend with around 8% v/v of alcohol concentration was selected to carry out the experimental work with an injection system and with busses and construction machinery. This work points out the most important advantages and disadvantages of the use of bioethanol blended with diesel fuel. The most important strength is the potential of these fuel blends for reducing particle matter without penalty in nitrogen oxides emissions. As main weakness can be cited the need for adding a stabilizer additive which restricts the desirable increase of bioethanol content in the blend.

Keywords: diesel, engine, vehicles, transient operation, emissions

1. Introduction

The interest of using renewable fuels started at the beginning of the 21st century due mainly to: the reduction of fossil fuels consumption and the decrease of carbon footprint of the combustion products. Although biodiesel is the biofuel most widely used in compression engines, bioethanol became in a competitor since its composition presents an oxygen concentration three times higher than biodiesel. Additionally, the production cost is competitive with fossil fuel because it can be produced locally through fermentation of sugar derived from corn or cellulosic biomass [1].

Bioethanol can be used as fuel in diesel engines by means three ways:

- Dual fuel system. Apart from the traditional diesel injection system, an additional system for the injection of bioethanol is added which can be located in the intake air manifold [2] or in the cylinder but independently of the injector of diesel fuel [3, 4]. This configuration implies an additional cost but allows to work with a higher ethanol-diesel ratio.

- Fumigation. Bioethanol is atomized using a carburation in the intake air manifold and introduced together with air in the cylinder. The main advantage is that the evaporation of ethanol reduces the temperature in the cylinder (increasing the air density) could reach higher engine power [5, 6]. However, the concentration of alcohol is limited to avoid the "knocking" at high engine load and the "misfiring" at low engine load [7, 8].

- Bioethanol-diesel blends. No engine modifications are required for blends up to 30% of alcohol concentration, making bioethanol-diesel blends (named as e-diesel) the most common way to use this alcohol as fuel in compression ignition engines. The main drawbacks associated to these blends are the limited miscibility of bioethanol in diesel and their poor lubricity properties.

Reductions in terms of pollutant emissions are associated to the use of e-diesel blends. Ethanol's oxygen reduces the probability of rich-zone formation and promotes the oxidation of soot nuclei generated, reducing smoke, Particulate Matter (PM) and other pollutant emissions such as Total Hydrocarbons (THC) and Carbon Monoxide (CO). These benefits are even further than those obtained with biodiesel-diesel blends [9, 10].

The effect of e-diesel blends under steady state conditions with different ethanol concentrations has been widely studied, being the alcohol concentrations most used 5%, 7.7%, 10% and 15%. Most of the bibliography remark the reduction of smoke opacity and PM emissions with these blends [11–13]. However, real driving conditions are majority transient sequences and the effect of fuel under transient conditions could be different since the engine working characteristics are also quite different to the stationary ones. Literature related to the effect of e-diesel blends under transient operation was scarce. Giakoumis et al. [14] reviewed the effect of e-diesel (and butanol-blends) under transient operations i.e., acceleration, load increase, starting process and driving cycles, until 2013. In the introduction of this chapter, this literature is updated and enlarged with studies related to the influence of bioethanol on the durability or wear of different engine parts.

Regarding engine testing, Ahmed et al. [15] carried one of the first works out by where two e-diesel blends (with 10% and 15% of alcohol concentrations) were tested under Federal Test Procedure (FTP) transient cycle of a heavy-duty diesel engine. Reduction up to 41% of PM emissions were recorded, being the decrease proportional to the alcohol concentration, with no penalty in NOx emission (slightly decreases were obtained).

Armas et al. [16] started to evaluate the effect of bioethanol testing an E10 blend under different transient sequences. Subsequently, this research studied the effect of the same percentage of ethanol, and a butanol-diesel blend (But16) under two different engine working conditions: the starting process (under warm and hot temperatures) and reproducing the New European Driving Cycle (NEDC) [17, 18]. Tests were carried out in a test bench with a light-duty diesel engine. Reductions in smoke opacity, PM and CO emissions were reported with e-diesel blend except during the cold starting process, where the high enthalpy of vaporization and low cetane number of bioethanol favor delayed premixed combustion process, increasing PM, NOx and THC emissions under this engine working sequence although the oxygen presence in the fuel.

The aim of the study of van Niekerk et al. [19] was investigating the effect of ethanol-biodiesel-diesel blends tested over the World Harmonized Light Vehicle Test Procedure (WLTP) using an ignition compression engine. Seven binary and ternary blends were selected by means a Design of Experiments (DoE). Pollutant emissions were measured, and different statistical models were used to describe trends obtained. Authors concluded that the ternary blend B2E9, considered as the optimum, showed reductions in CO emissions of 34%, in NOx emissions of 10% and 21% in CO_2 emissions compared to diesel.

Regarding vehicles testing, emissions tests were carried out by Randazzo et al. [20] with a light-duty vehicle positioned over a chassis dynamometer under NEDC conditions. In this work, bioethanol was added (at very low concentration, 2% and 5%) of two blends biodiesel-diesel with 20% of biodiesel concentration. Authors concluded that the use of bioethanol reduced NOx and CO_2 emissions but, surprisingly, it was unfavorable for CO, THC and PM emissions. Although the higher oxygen concentration of bioethanol (compared to biodiesel fuel), its lower cetane was considered as a factor that favored the PM emissions.

Two busses used for urban transportation were tested by Mata et al. [21] using a bioethanol-diesel blend (7.7% of alcohol) in two cities at different altitudes. Notable reductions in particle concentrations were obtained with e-diesel blend while the trend of NOx emissions depended on the altitude. The engine of other bus (a Cummins B Series) was used to evaluate the effect of two e-diesel blends (10% and 15%) under the 8-mode AVL test cycle conditions [22]. PM emissions decreased 20% and 30% with e-diesel blends but no noticeable effect on the emission of NOx and even small increases in THC and CO were observed.

Regarding non-road engines or vehicles testing (machinery), few works were found about the use of bioethanol as fuel in construction machinery vehicle. Armas et al. [23] evaluated the effect of an e-diesel blend with a 7.7% of ethanol in a vibration roller under four operation engine conditions: engine start, idle, circulation and work. Significant reductions of smoke opacity were obtained during transient engine operation.

Three non-road heavy-duty diesel engines were used in the work of Merrit et al. [24] where three e-diesel blends (E7.7, E10 and E15) were used. The FTP smoke test was reproduced, results being classified into acceleration, lugging, and peak modes. Smoke and particulate matter emissions decreased (up to 20%) as ethanol concentration increased. CO emissions were also lower than those of diesel base fuel while NOx concentration was similar.

As it was commented previously, the main drawback of e-diesel blends is the limited miscibility of both fuels which requires knowing conditions where these blends are stable. Bioethanol's miscibility in diesel mainly depends on three parameters: water content (hydrous ethanol), temperature and ethanol concentration [25]. The presence of water in ethanol, the low temperatures and the high ethanol concentration difficult the miscibility between both hydrocarbons [25–27].

To ensure the miscibility between bioethanol and diesel, especially at low temperatures, different additives are usually added: alcohols (butanol, pentanol, octanol, dodecanol) [28, 29], ethers, and different co-solvents [26, 30] or emulsifiers [31, 32] whose composition is, sometimes, not published. One of the most common additives for bioethanol-diesel blends is biodiesel fuel which, in addition to its renewable origin, contributes to improve the cetane number and lubricity of e-diesel blends [33, 34]. Different authors evaluated the stability of bioethanol-biodiesel-diesel (named as e-b-diesel) blends at different temperatures, indicating the different instability areas [9, 35].

Apart from the limited miscibility, easily resolved by the incorporation of additives, the poor lubricity properties of bioethanol could imply wear of the fuel injection system and other engine components. Two types of investigations have been carried out in relation to this topic:

- Wear of metallic materials. Wear Scar Mean Diameter (WSMD) of different e-diesel and e-b-diesel blends was evaluated in the work of Lapuerta et al. [36].

The increase of WSMD was not linear with the increase of ethanol concentration. Durability tests of fuel injection system using several experimental techniques were carried out by Armas et al. [37]. A E7.7 blend showed similar effect on durability of the injection pump parts than that the diesel fuel but a reduction (around 20%) of nozzle hole effective sections was produced with the blend. In the rest of literature, wear of the injection system parts with bioethanol-diesel blends is observed [38, 39].

- Material degradation/oxidation. Some metals (zinc, plumb, copper, aluminum) and some plastics (rubber and polyvinylchloride, PVC) are oxidized when contacting with bioethanol [40]. To solve this drawback, some authors suggest the use of stainless steel as a substitute for metals (aluminum, magnesium and brass) [41, 42] and, in other work, high-density polyethylene is used instead of rubber or other plastic materials in [40].

The aim of this chapter is to show the main results of several works about the effect of bioethanol-diesel blends on the durability of fuel injection system and on the performance and emissions in diesel engines or vehicles under transient working conditions.

2. Experimental installations, procedures and fuel blends characteristics

2.1 Engine testing

With the objective to evaluate the effects of bioethanol blended with diesel fuel, under transient operation, previously to the direct use in vehicles, an experimental study with an engine mounted on a test bench is recommendable. In this case, **Figure 1** shows a general sketch of the experimental installation used for testing the engine under discrete transient sequences. A turbocharged intercooled, direct injection Diesel engine, 4-cylinder, 4-stroke, typically equipped in European light duty vehicles, was employed as experimental unit. In **Table 1**, main characteristics of the engine tested are listed.

The engine was coupled to an asynchronous machine Schenck Dynas3 LI 250 (operating as a dynamometer). The dynamometer control system allowed measuring, controlling, and registering the engine speed, accelerator position and effective

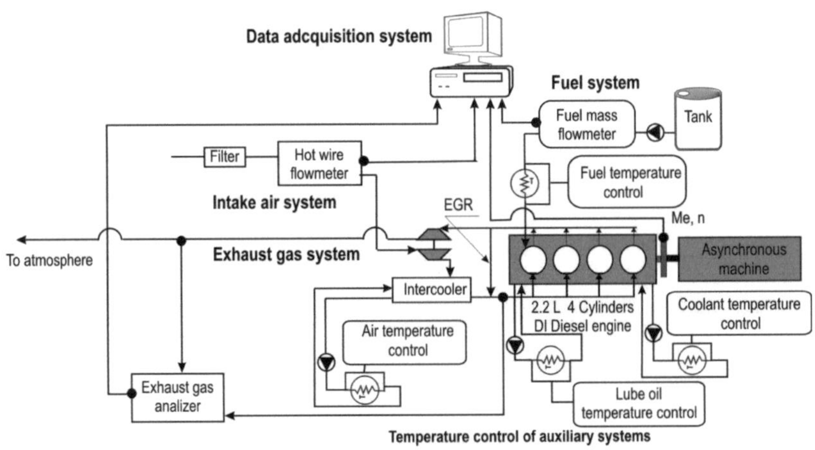

Figure 1.
Sketch of the experimental installation for engine testing.

Parameter	Value
Compression ratio	18:1
Swept volume (L)	2.2
Stroke (m)	0.94
Bore (m)	0.865
Cylinders arrangement	4, in line
Maximum rated torque (at 2000 min^{-1}) (Nm)	237.4
Maximum rated power (at 4000 min^{-1}) (kW)	85.2 kW
Fuel injection system	VP44 pump, electronically controlled
Main injection pressures (bar) at full load (at 2000 min^{-1}) at idle (750 min^{-1})	1100 200

Table 1.
Main characteristics of the engine tested.

torque. Instantaneous fuel consumption was registered by means of a PLU 401/116H flow meter, while the air mass flow rate was measured by means of a hot-wire flow meter Siemens 5WK9628, previously calibrated in the 0–718 kg/h range with an accuracy of 2% around the measured value.

The Exhaust Gas Recirculation (EGR) ratio was calculated by comparison of the CO_2 gas concentration at the inlet and at the exhaust manifolds using an infrared absorption gas analyzer Environnement MIR2*M*. *NOx* emissions were measured by means of a chemiluminescence analyzer Environnement TOPAZE. In both cases (CO_2 and NOx) the acquisition frequency was 3 Hz. The smoke opacity of the exhaust gas (as characteristic parameter of particulate matter emission) is proportional to the total light extinction across the exhaust gas stream. Using a partial flow opacimeter AVL 439 with a frequency of 10 Hz, the smoke opacity was registered. Instantaneous engine parameters were registered by means of a Yokogawa OR1400 data acquisition system. This system has the possibility for data registering in a range from 1 Hz to 100 kHz. Finally, the sampling frequency of gaseous emissions, test bench parameters and the smoke opacity was 10 Hz.

For assessing the effect of changes in engine load on the smoke opacity and NOx emissions, two discrete transient sequences were tested and compared. In each sequence both, the initial and final conditions of effective torque and engine speed, were the same independently of the fuel tested.

Figure 2 shows these two transient sequences, both with effective torque (Me) increase at relative low engine speed (n) and denoted as A_M-A_F and A-A'. The aim of this methodology was to evaluate the smoke opacity and NOx emissions behavior under two of the most common transient sequences with fuel consumption demand in any type of vehicles. Under A_M-A_F sequence, the engine was driven from motored to fired conditions. This test tries to simulate a vehicle acceleration from zero fuel delivery to a given load condition. Meanwhile A-A' is the typical transient sequence with load increase at constant speed (for example, under a slope positive change in a road profile, remaining constant the vehicle velocity without gear changes). Under both cases, the engine control unit (ECU), tuned by manufacturer, controlled the EGR valve opening.

Initial and final operating modes of the sequences shown in **Figure 2** and **Table 2** were chosen due to their operation representativeness of that zone on the engine torque-engine speed map (under low load and engine speed) where the pollutant emissions restrictions are very important (typical of urban driving conditions).

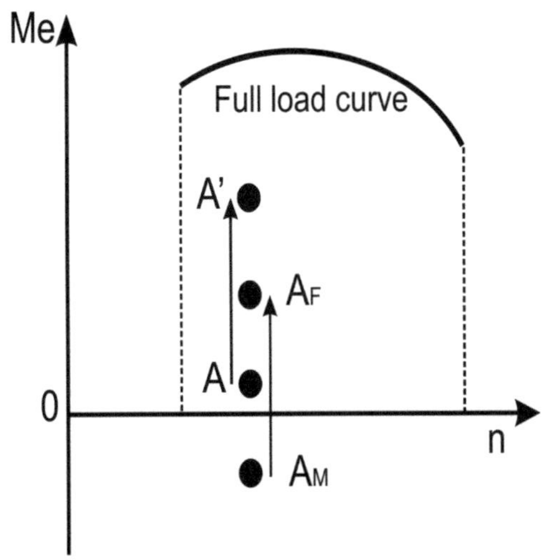

Figure 2.
Engine transient sequences tested.

Transient sequence	Engine speed (min^{-1})	Effective torque (Nm)	
		Start	End
A_m-A_f	1661	−40	55
A-A’		26	90

Table 2.
Steady state operating conditions before and after each transient sequence.

Under engine transient testing, a low Sulfur diesel fuel was used as reference (denoted as Ref). A 10% v/v blend of bioethanol with this diesel fuel, without stabilizing additive, was denoted as E10. This blend was selected because the ambient temperature was higher than 25°C. Under these thermal conditions and, during relative short time period, the blend is stable without additive since Bioethanol used has a purity of 99.94%. The bioethanol concentration of the tested blend was defined attending to the results of previous studies [26]. The main properties, either measured or calculated, of the tested fuels are presented in **Table 3**.

2.2 Parts durability testing

Before the extension of the work for assessing the benefits of bioethanol-diesel blends in vehicle captive fleets, two works must be done: a study about the miscibility and stability of the blends with different bioethanol concentrations [26] and the study of the effect of the fuel blend on the integrity of some parts of the engine injection system [37]. Both works are essential for verifying the lubricant capacity of the fuel blend under real driving conditions. This sub-section describes, as example, part of the work done with some pieces of a modern injection system.

The experimental unit used was a Bosch high-pressure injection pump. Normally this pump operates connected to a common rail and this one connected to a fuel injector, electronically controlled. This model of experimental unit equips great number of different models of Mercedes Benz light duty cars. This system also equips great part of diesel engines, with around 2 liters of cylinder displacement, commercialized in Europe. The main characteristics of the injection system studied in this work are shown in **Table 4**.

Property	Fuel		
	Ref	E10	E
Density (kg/m^3)[a]	833.5	828	792
Kinematic viscosity (cSt)[b]	2.79	2.13	1.13
Gross heating value (MJ/kg)	45.89	43.48	28.05
Low heating value (MJ/kg)[c]	42.84	40.44	25.18
Low heating value (MJ/L)[c]	35.70	33.49	19.94
% C (in weight)	85.23	82.08	52.17
% H (in weight)	13.92	13.83	13.04
% O (in weight)	0.74	3.98	34.78
% S (in weight)	0.026	0.024	0
Molecular weight	206.9[d]	155.2[e]	46.06[e]
Stoichiometric fuel-air ratio[e]	1/14.60	1/14.07	1/9.01
Distillation			
Initial Boiling Point (°C)	182	78	78
T10 (°C)	204	79	
T50 (°C)	266	257	
T90 (°C)	348	347	

[a]*Measured at 15 °C.*
[b]*Measured at 40°C.*
[c]*Calculated from composition and gross heating value.*
[d]*Calculated with software Aspen-Advisor.*
[e]*Calculated from composition.*

Table 3.
Fuel properties used under engine testing.

Injection system characteristic	Value (unit) and/or information
Pump manufacturer	Bosch
Pump type	first generation with pistons
Common rail model	270 CDI MB/4 injectors
Common rail length (m)	0.55
Pump-common rail line length (m)	0.40
Common rail-injector line length (m)	0.15
Inner diameter of lines (m)	0.002

Table 4.
Main characteristics of the injection system tested.

Figure 3 shows the scheme of the hydraulic circuit of the experimental installation used for simulating a long-term work of an injection system under demanding operating mode. The injection system worked on an injection test bench (model Asia Diesel). In dark black, the circuit (without modifications) for testing diesel fuel is presented. Those parts modified for testing bioethanol-diesel blends are presented in light gray (valve 2 and tank caps). Those parts were added to remain the water content in the range established by the European fuel quality standard EN-590 and for avoiding the ethanol evaporation to the atmosphere. In addition,

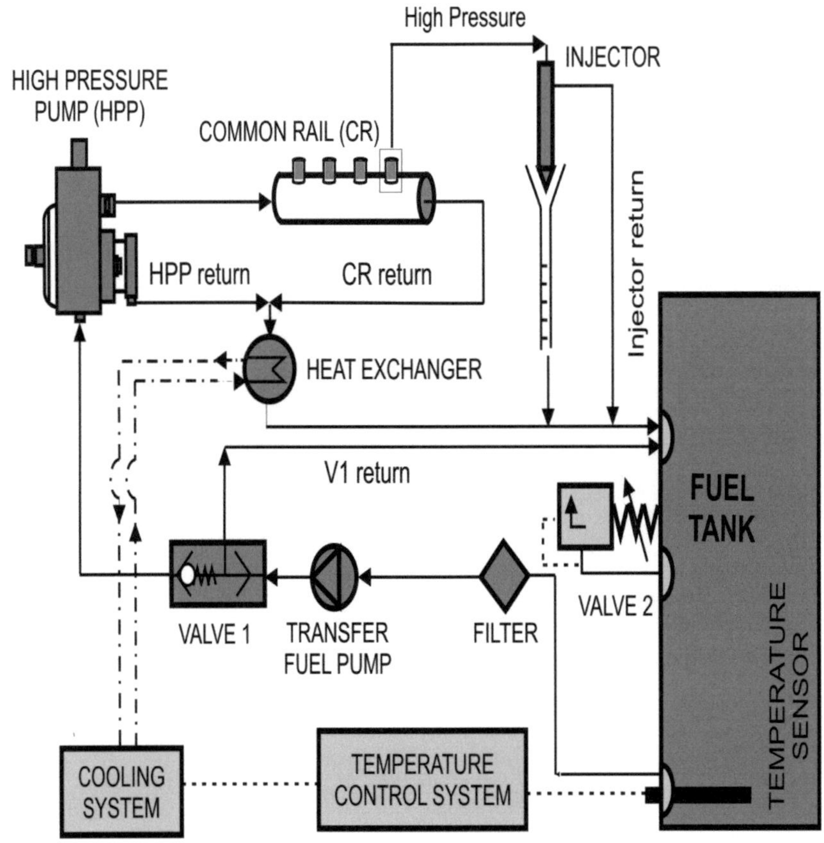

Figure 3.
Scheme of the hydraulic circuit of the test bench for simulating long-term work with fuels tested.

the fuel tank was instrumented with a pressure and/or vacuum safety valve, to prevent the pressure variation inside the fuel tank. These modifications were also needed in the fuel tanks of the vehicles tested.

Table 5 shows the test conditions for simulating reproduced by the test bench during the long-term work of a new injection system used with each fuel tested.

Before and after the long-term operation with each fuel, the total fuel delivery was determined using diesel fuel and those conditions presented in **Table 5**. Selected test conditions (600 work hours at 2500 min^{-1} and 1500 bar of injection pressure) are equivalent to the driving of a light duty vehicle during more than 120,000 km.

For characterizing, the effect of fuels on the integrity of some parts of the injection system, different techniques can be used. Among these techniques are the

Parameter	Unit	Value
Rotation speed	min^{-1}	2500
Injection pressure	bar	1500
Injection time	ms	1
Fuel temperature	°C	40
Operation time	hours/day	10–12
Total test time	hours	600

Table 5.
Test conditions for simulating reproduced by the test bench.

following: measurement of the fuel delivery, characterization of the roughness surface of elements, observation of the surface microstructure, weigh of elements and measurement of the nozzle geometry inner shape obtained by casting silicone, determined from images of a Scanning Electronic Microscope (SEM). In this chapter, the effect of E8 fuel blend on the surface roughness of the drive shaft of the fuel high-pressure pump (**Figure 4a**, surfaces Ds1, Ds2 and Ds3), the geometry of the nozzle (**Figure 4b**, diameters along the nozzle hole in three positions d1, d2, d3) and the total fuel delivery is compared. Comparison was done, before and after 600 work hours, with diesel fuel.

For determining the effect of the long-term work on the surface roughness a tester Hommel Werke, model T500, was used. This tester has capacity to determine different parameters of surface roughness, along 10 mm of the sampled surface, with 100 microns amplitude and with a precision of 0.01 μm. Diameters along the nozzle hole were determined from images obtained by means of a scanning electronic microscope Philips model XL30 of the hole inner shape obtained by casting silicone [43].

This part of the work and the work done with both busses and with construction machine, as will be explain later, was carried out with a different bioethanol-diesel fuel blend. In these two cases, a low Sulfur diesel fuel was used as reference (denoted as Ref). A 7.7% v/v bioethanol-fuel blend, with 0.62% v/v of a confidential stabilizing additive, was used and denoted as E8. **Table 6** shows properties of fuels used in this part of the work. Bioethanol purity was 99.94%. Since the additive composition was not known, all E8 fuel characteristics, dependent on the fuel composition, were determined only from diesel and bioethanol fuels.

2.3 Vehicle testing. Urban busses

Two similar busses used for urban transportation were tested. One of the vehicles was a Euro II IVECO Europolis 915 city bus (denoted as IV), around 10 T in weight, equipped with a 6-cylinder, with around 6 L of displacement, direct injection, turbocharged, heavy-duty diesel engine. This bus was equipped with an automatic ZF transmission. The bus length is 9 m. The other vehicle was a Euro II Renault city bus (denoted as RE), around 13 T in weight, equipped with a 6-cylinder, with around 8 L of displacement, also turbocharged, direct injection, heavy-duty diesel engine. This bus was equipped with an automatic Voith transmission and its length is 10 m. Both busses were equipped with injection systems including injection pumps electronically controlled. Both busses operate with single

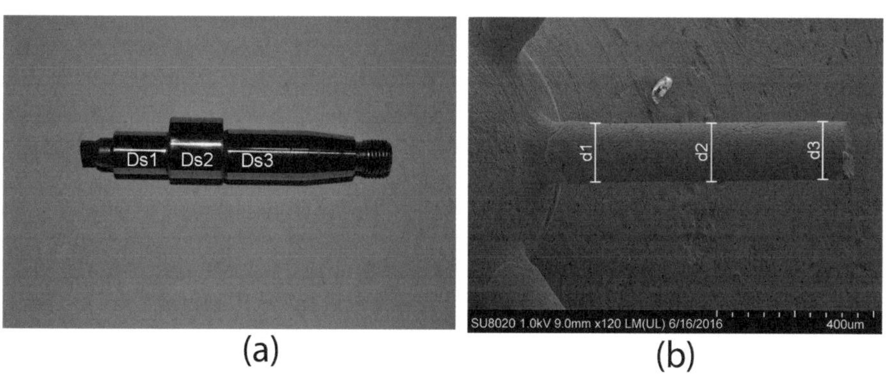

Figure 4.
a) Surfaces of the drive shaft of the high-pressure pump studied and (b) example of a SEM image from the silicone mold of a hole of the injector nozzle.

Property	Fuel		
	Ref	E8	E
Density (kg/m^3)[a]	834.9	831	792
Kinematic viscosity (cSt)[b]	2.72	2.41	1.13
Gross heating value (MJ/kg)	45.54	43.82	28.05
Low heating value (MJ/kg)[c]	42.58	40.86	25.22
Low heating value (MJ/L)[c]	35.55	33.95	19.97
% C (in weight)	86.13	83.63	52.14
% H (in weight)	13.87	13.82	13.13
% O (in weight)	0	2.55	34.73
% S (in weight)	0.0034	0	0
Molecular weight	211.7[d]	167.5[e]	46.06[e]
Stoichiometric fuel-air ratio[e]	1/14.67	1/14.25	1/9.00
Distillation			
Initial boiling point (°C)	172	78	78
T10 (°C)	211	178	
T50 (°C)	270	256	
T90 (°C)	340	337	

[a]*Measured at 15 °C.*
[b]*Measured at 40°C.*
[c]*Calculated from composition and gross heating value.*
[d]*Calculated with software Aspen-Advisor.*
[e]*Calculated from composition.*

Table 6.
Properties of fuels used in parts durability testing and in busses and construction machinery testing.

injection and advanced Start of Injection (SoI) and without Exhaust Gas Recirculation (EGR). Both vehicles, with similar mileage, were not equipped with after-treatment devices (Diesel Oxidizer Catalyst (DOC) nor Diesel Particulate Filter (DPF)). This configuration of the exhaust systems allowed studying the effect of fuels on pollutant emissions without influence of after-treatment devices.

As **Figure 5** shows, both busses and the construction machine were instrumented with a HORIBA OBS 1300 gas analyzer and with a TSI Engine Exhaust Particle Sizer (EEPS) spectrometer. The first of them includes sensors for measuring and registering the relative fuel-air ratio and ambient conditions (temperature, pressure and humidity). In both cases, vehicle velocity and its position were determined by means of a Global Positioning System (GPS) and an inertial sensor. Although the HORIBA OBS 1300 includes different measurement modules, only the MEXA 720 NOx with a zirconia sensor for nitrogen oxides measuring was used. For characterizing particulate matter, an EEPS spectrometer was used to measure particle size distributions under transient vehicle operation. **Figure 5** also shows, the EEPS spectrometer coupled to a Rotating Disk (RD) diluter model MD19-2E, as primary dilutor, and an air supply-thermal conditioner model ASET15–1 with an evaporating tube, as secondary dilutor.

At Ciudad Real city, the tests were carried out in a route with 14.5 km length (around one hour in time). This route mostly runs throughout the city center, characterized by a low mean velocity. However, it includes some stretches of

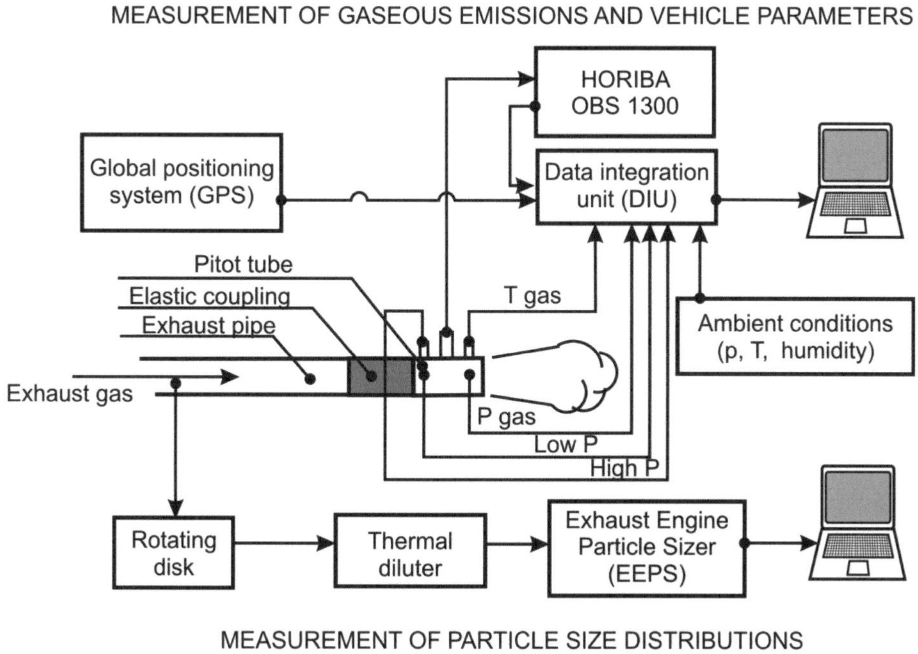

Figure 5.
Sketch of the experimental installation for vehicle testing (busses and construction machine).

medium speed (around 50 km/h). The altitude profile is almost constant during the entire route (~650 m above the sea mean level). The route has 67 stops for passengers, uniformly distributed along the trajectory and it includes 31 traffic lights.

At Sevilla city, the tests were carried out in route with 13 km length (also one hour in time). The route runs mostly throughout the city center with similar mean velocity as occurred at Ciudad Real city. The altitude profile is almost constant during the entire route (~10 m over the sea level). The route has 35 stops for passengers, uniformly distributed along the trajectory and it includes 21 traffic lights.

The comparison of the results obtained from real driving operation of test vehicles is relatively complex. By this reason, results presented in this chapter were processed following the methodology presented in [44]. In this case, comparison between busses and fuels, based on the obtained mean values from categories, will be presented below. Cycle is the whole data from the beginning to the end of the route while category is that part of the cycle with similar variation of the most important operating parameters (relative fuel-air ratio, denoted as Fr and vehicle velocity, denoted as Vv).

A great number of different categories such as accelerations, decelerations, idle, etc. composes one cycle [44]. In this text only the analysis of part of the events included within the acceleration category is presented. Within the acceleration category, there are three possible situations: a) acceleration coming from idle, b) acceleration coming from deceleration without fuel consumption and c) acceleration coming from deceleration with fuel consumption. As example, **Figure 6** shows vehicle velocity and fuel-air ratio profiles, registered during part of a cycle.

Four categories, idle (I), acceleration (A), deceleration with fuel consumption (DwF) and deceleration without fuel consumption (D) [21] were selected as example for this work. In this chapter, as example, only the comparison under acceleration coming from idle sequence is presented.

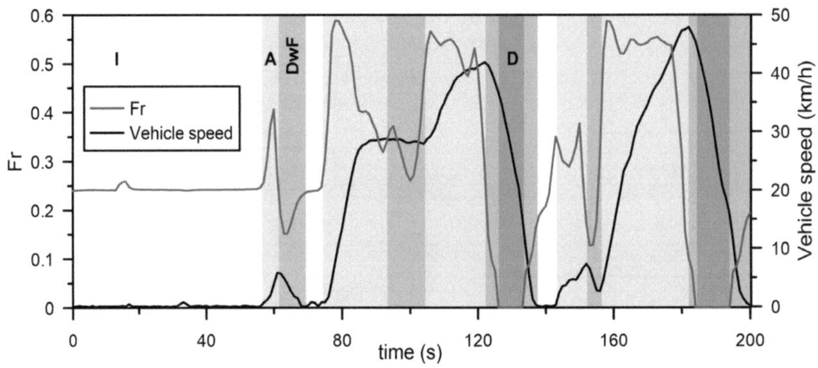

Figure 6.
Vehicle velocity and relative fuel-air ratio profiles from part of the bus cycle.

2.4 Vehicle testing. Construction machinery

As experimental unit was used a vibrating roller Lebrero model Rahile 155TT. This machine equips a direct injection Diesel engine, 6-cylinder, 4-stroke turbo-charged, typically used during roads and railways construction. **Table** 7 shows the most important characteristics of its engine.

This machine was instrumented with a HORIBA OBS 1300 system as **Figure** 7 shows. Its connection to the exhaust pipe was similar to that used in public busses. However, the smoke opacity (indicative of the particulate matter emitted), which ranges from zero to 100% and proportional to the total light extinction across the exhaust gas stream, was determined by a smoke opacimeter Wager 6500. This signal was also registered by the OBS system.

The comparative analysis of pollutant emissions registered under the operation of a construction machine is complex. As occurred with public busses and regarding the engine work, the work done by the machine can be classified into five categories [23]. These categories are engine start on (denoted as S), idle (denoted as I), circulation (denoted as C) and work (denoted as W).

This categorization allows the daily machine operation without disturbance on its own work. This machine, during the tests, worked under real operating conditions during the construction of a railway. The best option is to register engine parameters and pollutant emissions in different occasions during the selected

Parameter	Value
Model	Deutz FL6913 turbocharged
Rated power highly intermittent at 2500 min^{-1} (kW)	109
Rated torque at 1700 min^{-1} (Nm)	510
Compression ratio	15.5:1
Displacement (L)	6.128
Cylinder arrangement	6, in line
Bore (m)	0.102
Stroke (m)	0.125
Fuel injection system	Bosch in line injection pump

Table 7.
Main engine characteristics of the construction machine.

Figure 7.
View of the experimental installation located on the construction machine.

operation sequences. In result and discussion section, results related to C and W sequences are presented as example.

3. Results

3.1 Engine testing

Time evolution of the smoke opacity (characteristic parameter of particulate matter emission) and NOx concentration, under transient sequence A_M-A_F, is presented in **Figure 8**. This behavior can be explained because both relative fuel-air and EGR ratios were zero at the beginning of the transition avoiding, this way, soot depositions on the internal of the manifold walls and thus any further blowing effect. Great reduction of smoke opacity is observed when E10 blend is used compared to reference fuel. The increase of the oxygen content of the fuel blend and the reduction of the aromatic content explain this result. Additionally, as **Figure 9** shows, the EGR ratio at the end of the transient sequence with the E10 blend, would hardly reach that of reference fuel. This also would leads to a greater availability of oxygen at the end of the transition, contributing to the smoke opacity reduction registered.

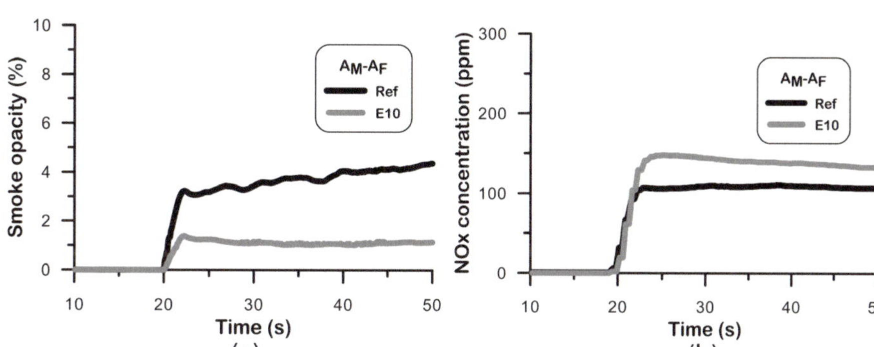

Figure 8.
Time evolution of the smoke opacity (a) and NOx concentration (b) during the A_M-A_F transition.

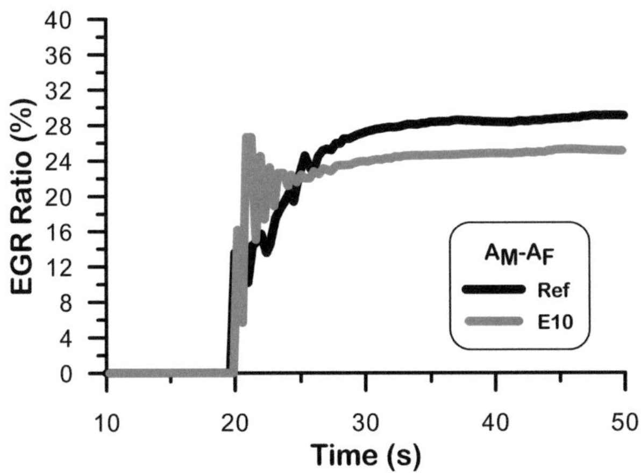

Figure 9.
Time evolution of EGR ratio during the A_M-A_F transition.

However, the slight increase of NOx concentration registered along the transition with E10 blend (see **Figure 8b**) can be explained by the initial oscillations of the EGR ratio (due to the sensibility of the EGR valve opening) presented in **Figure 9**. It is important to take into account two factors: i) manufacturer tuned the engine (included the EGR valve opening) with diesel fuel and ii) under a typical diesel combustion process, increases in EGR ratio generally produce decreases of NOx concentration but increases of smoke opacity, this last as characteristic parameter of particulate matter.

Figure 10 presents the smoke opacity and NOx concentration time evolution along the transition A-A' registered with both fuels. As **Figure 10** shows, the shape of the smoke opacity and NOx concentration curves during their time evolution are similar between both fuels.

This shape is consistent with the shape of the relative fuel-air ratio (related to the stoichiometric fuel-air ratio) and EGR ratios as **Figure 11** presents.

In this figure, as example, is presented the time evolution of both parameters registered with both fuels. It is important to remember that the engine is forced to reach the same torque value at the beginning and at the end of the transition.

The sudden increase of the relative fuel-air ratio is caused by the delayed response of the turbocharger. As consequence, this produces a time lag between the fresh inlet airflow and the increase of fuel delivery demanded to reach the torque target at the end of transition. Additionally, other contributor factor of the smoke

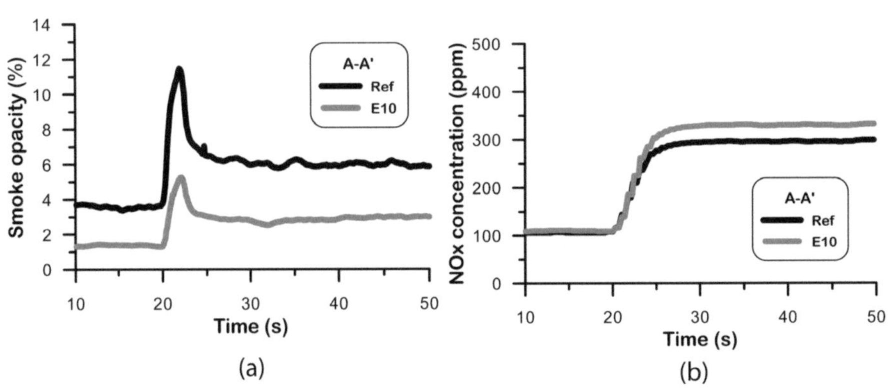

Figure 10.
Time evolution of the smoke opacity (a) and NOx concentration (b) during the A-A' transition.

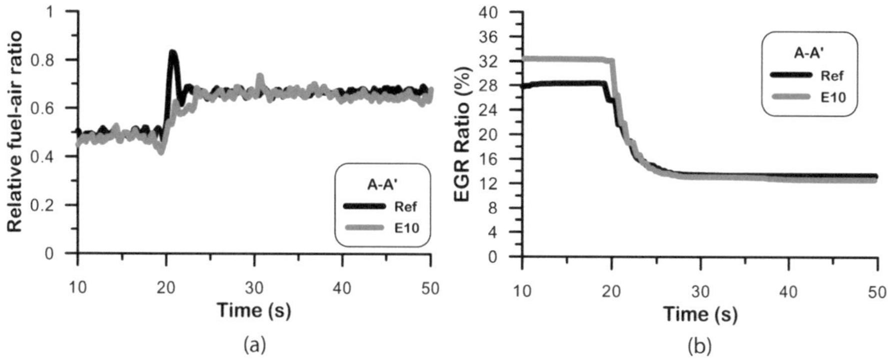

Figure 11.
Time evolution of relative fuel-air ratio (a) and EGR ratio (b) during the A-A' transition.

opacity peaks is the release of soot deposited on the pipe walls during the steady operation previously to the transition, which could be blown out as consequence of the increased exhaust gas flow with higher thermal conditions.

As can be seen in **Figure 10a**, the smoke opacity is clearly lower with E10 blend. The lower viscosity and higher volatility of E10 fuel could favor the fuel spray atomization and its evaporation, just during the transition. Effectively, the reduced content of aromatic compounds and the higher oxygen content of the E10 blend follows being the most important contribution to the reduced smoke peak.

Regarding NOx concentration presented in **Figures 8b** and **10b**, different properties of the bioethanol-blend cause opposite effects on this pollutant emission. First, the low adiabatic flame temperature and the high vaporization heat of the E10 blend reduce the combustion temperature and the NO formation [45]. Second, a low cetane number (leading to longer delay time), the high oxygen content, a fast combustion velocity and an advanced fuel injection (this being a response of the control unit for compensating the longer injection process due to the reduced heating value), are those factors which favor high combustion temperature and NO formation [45–48]. According to these two groups of factors and depending on the engine operating mode and the type of engine both increases and decreases of NOx with bioethanol-diesel blends have been reported [45–48].

These results indicate the great potential to obtain important reductions of smoke opacity without penalty of NOx emissions by tuning the engine with a bioethanol-diesel blend.

3.2 Parts durability testing

For the surface roughness analysis, two parameters were used: the arithmetic mean roughness value (Ra) and the mean peak to valley height (Rz), both defined by the DIN4777 and ISO4287 standards. The mean roughness (Ra) is defined as the arithmetic mean of the profile deviation of the filtered roughness profile from the mean line (ML) within the measuring length (L), according to Eqs. (1) and (2) and the representation shown in **Figure 12**.

$$R_a = \frac{1}{L}\int_0^L |y|dx \quad (1)$$

$$R_a = \frac{\sum^i Areas\ A_i + \sum^j Areas\ A_j}{L} \quad (2)$$

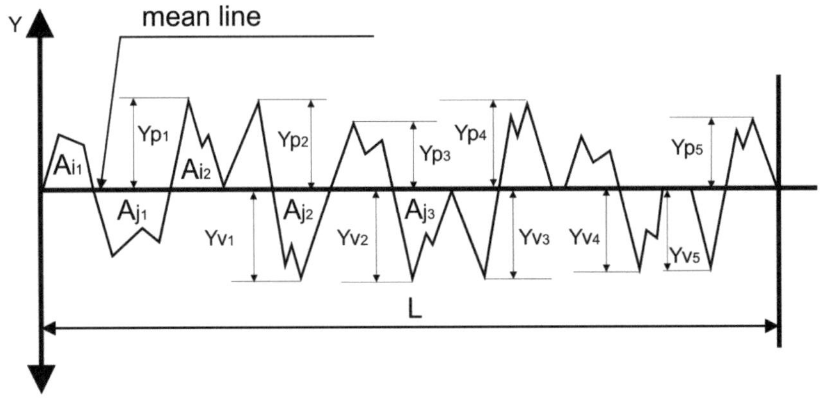

Figure 12.
Sketch of the surface roughness with peaks and valleys.

The mean peak to valley height (Rz) is defined as arithmetic mean from the peak to valley heights y_1 to y_5 of five successive sampling lengths in the filtered roughness profile, according to Eq. (3).

$$R_z = \frac{\sum_{i=1}^{i=5} y_{p_i} + \sum_{j=1}^{j=5} y_{v_j}}{5} \tag{3}$$

The absolute differences ΔRa and ΔRz were calculated by means of the Eqs. (4) and (5) in order to determine the effect on surface roughness of each fuel tested before and after of the long-term work. These differences were obtained between the final and initial mean values of the Ra and Rz parameters of each zone of the drive shaft shown in **Figure 4a**.

$$\Delta Ra = Ra(600h) - Ra(0h) \tag{4}$$

$$\Delta Rz = Rz(600h) - Rz(0h) \tag{5}$$

Ra and Rz parameters measured on the drive shaft surfaces Ds1, Ds2 and Ds3, with both fuels, are presented in **Figure 4a**. In addition, the 95% confidence interval using three measurements at each point is indicated. According to the results presented in **Figure 13**, the greatest Ra and Rz parameters values were calculated in that surface denoted as Ds3 and this occurred independently of the fuel tested.

The surface roughness measured on this zone was more pronounced than those measured on Ds1 and Ds2 surfaces were. The zone Ds3 is located on the most loaded section surface of the drive shaft. The measured surface roughness fits in the grade N4 to N2 (N is the surface finish grade as per the standard ISO1302). This classification corresponds to machined surfaces derived from lapping and/or finishing turning operations. It is important to highlight that if the surface roughness of two contact surfaces drastically decreases (below 0.025 mm), it could lead to the blockage of the elements due to surface adherence (lapping effect). In the surface Ds3), the flange transmits a great force to the drive shaft, and this action leads to a high friction value. The values calculated of the parameters Ra and Rz for the surfaces Ds1, Ds2 and Ds3 were similar, prior to and after the durability test, independently of the type of fuel used. **Figure 14** shows the maximum ΔRa and ΔRz values that were measured on each point along the drive shaft.

Results presented in **Figures 13** and **14** indicate similar effect of both fuels tested along the drive shaft. The lower lubricity of the E8 fuel blend produces a negligible effect on the surface under a long-term work equivalent to more than 120,000 km.

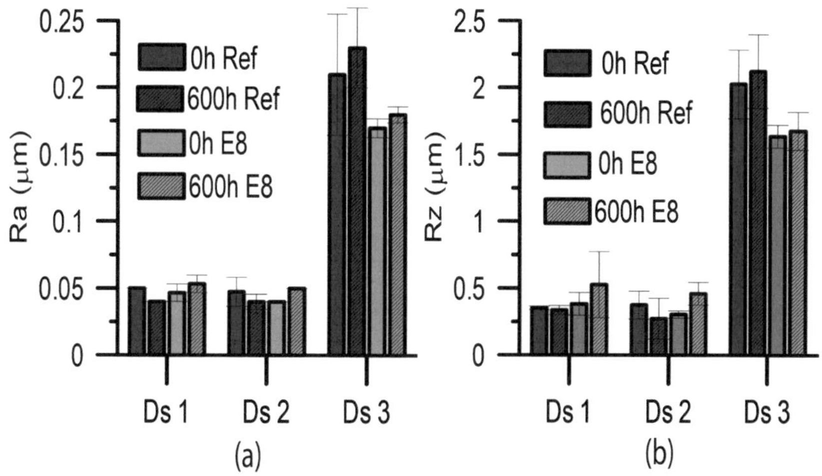

Figure 13.
Ra and Rz parameters calculated for surfaces Ds1, Ds2 and Ds3 of the drive shaft.

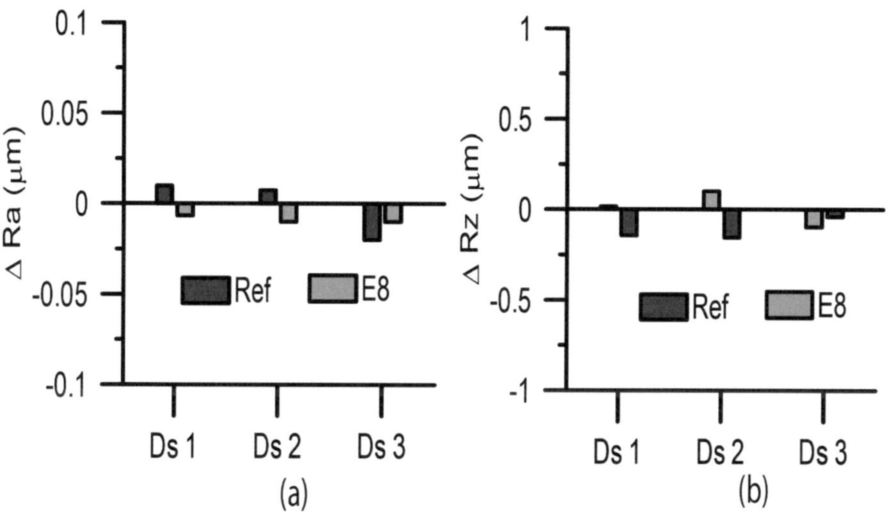

Figure 14.
Maximum ΔRa (a) and ΔRz (b) values determined from each surface of the drive shaft.

Table 8 presents diameters (d1, d2 and d3) measured from the silicone casting images taken from two randomly selected nozzle holes used with each fuel, before and after the tests. This table also shows the relative difference (ΔAd) of the

Fuel	Hole	d_A (μm) (600 h)			Mean d_A (μm)	d_B (μm) (0 h)			Mean d_B (μm)	ΔAd (%)
		d_1	d_2	d_3		d_1	d_2	d_3		
Ref	1	192.8	192.8	193.0	192.9	192.8	192.8	193.1	192.9	0
	2	193.0	192.8	193.2	193.0	192.8	192.7	193.0	192.8	−0.2
E8	1	200.7	200.6	199.8	200.4	175.4	182.0	184.0	180.5	−18.9
	2	201.9	200.9	202.9	201.6	172.5	178.2	179.3	176.7	−23.2

Table 8.
Geometrical characterization of sections of nozzle holes.

effective sections (A) before (denoted as B) and after (denoted as A) the tests, calculated from mean diameters d_B and d_A, respectively.

Differences obtained, section by section, among diameters of a same nozzle used with each fuel is lower than 1.5%. This value is within the dispersion range of Bosch nozzles. Reductions of the effective sections of 18.9% and 23.2% respectively were obtained after using E8 fuel blend. These reductions can be explained by a probable sedimentation and/or oxidation along nozzle holes. This explanation could be justified because of the test bench was not completely hermetic. This situation provoked a slight reduction of ethanol concentration (around 0.2% v/v). Additionally, the water concentration increased in the blend from 243 ppm to 460 ppm, after 600 h. After the test, the experimental installation was checked. One point without a correct sealed was detected: the nozzle tip. This fact could produce the contact between the ambient air with the nozzle tip. The air oxygen could produce a slight oxidation along the holes, causing the reduction of their effective sections (**Table 9**).

The total fuel delivery determined before and after the tests with each fuel is shown in **Table 10**. This parameter varied within a narrow range of variation (around 3%), before and after the long-term work with diesel fuel. Contrary, the total fuel delivery decreased approximately 30% after the test with E8 fuel blend.

3.3 Vehicle testing. Urban busses

In this section, the time analysis of acceleration sequences (A) registered through the test cycles is presented. This analysis has been carried out by comparison of five seconds of those sequences with similar time profiles of the relative fuel-air ratio (Fr). **Figure 15** presents the average time evolution of Fr, Vv, NOx concentration and its mass flow rate from both busses and fuels tested respectively.

As can be seen in **Figure 15**, the Fr profiles between vehicles and fuels were similar enough for comparing the effects of fuels. NOx concentrations from E8 fuel were lower compared to diesel fuel when the IV bus was tested, and it was higher when the RE bus was tested. In both cases, the difference between average NOx concentrations produced by fuels, in both vehicles, was practically constant along the sequence.

Fuel	Fuel delivery (cm^3 per stroke)		
	Before	After	Relative difference %
Ref	6.0	6.2	+3
E8	6.1	4.3	−29.5

Table 9.
Total fuel delivery before and after the tests with both fuels.

Sequence	Parameter						
	Fuel	Fr	T_{amb} (°C)	T_{exh} (°C)	$\dot{V}_g$ (L/min)	$\dot{m}_{fuel}$ (g/s)	υ (km/h)
C	Ref	0.308	9	243	5444	1.221	7.6
	E8	0.312	0	238	4025	0.950	6.6
W	Ref	0.528	2	387	5905	1.750	6.7
	E8	0.471	3	348	4638	1.340	3.5

Table 10.
Mean values of engine and/or vehicle operating parameters obtained during the W and C sequences.

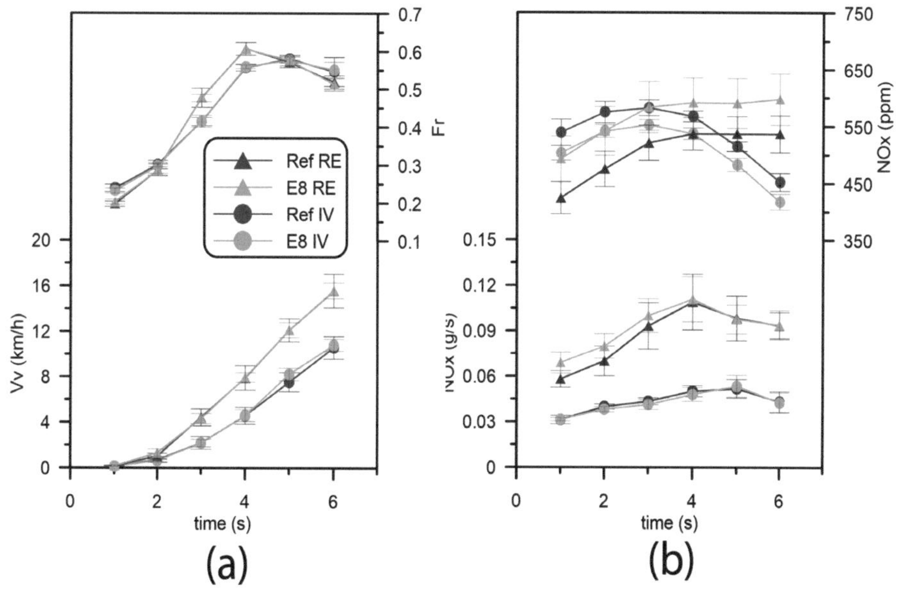

Figure 15.
Relative fuel-air ratio (Fr) and vehicle velocity (Vv) profiles (a) and NOx concentration and mass flow rate profiles (b).

Low NOx concentration produced by the E8 fuel blend when the IV bus was tested at Ciudad Real city, can be explained as follows: the higher enthalpy of vaporization and its cooling effect, combined with the lower oxygen availability at altitude, lead to a decrease in NOx concentration. This occurred even when the engine compression ratio of IV bus is higher than that of the RE bus. An opposite effect was registered the RE bus tests. The lower altitude of Seville leaded to a higher air oxygen availability during the combustion process that, together with the oxygen content of E8 blend, produced an increase of NOx concentration. This occurred even the lower engine compression ratio of the RE bus and the higher enthalpy of vaporization of E8 fuel blend both compared to the IV bus and the diesel fuel respectively.

Compared along the time analyzed and with each bus, NOx mass flow rates are similar between fuels. NOx mass flow rates differences between busses can be explained by the different engine displacement of busses. The RE engine is two liters larger than the IV engine. More displacement leads to more quantity of gas displaced by the engine.

Figure 16 shows time evolution of particle concentration emitted by busses with both fuels under the acceleration sequence. As shown in this figure, particle concentration evolution followed the time profiles of Fr (see **Figure 16a**).

Maximum particle concentration is reached around the 4th second as occurred with maximum Fr. Particle concentration increases proportionally respect to the Fr increase. As lower the oxygen concentration in the air as higher the particle concentration. Compared to diesel fuel, in both busses, particle concentration produced by E8 fuel blend is lower. Vehicles engines reproduce similar trend seen on the engine test bench. The lower quantity of aromatic compounds and the presence of molecular oxygen in E8 fuel blend lead to a cleaner combustion process and explain the obtained results.

3.4 Vehicle testing. Construction machinery

Average values of some characteristic parameters of the work done by the roller engine are listed in **Table 8**. These are indicative of the engine and/or vehicle

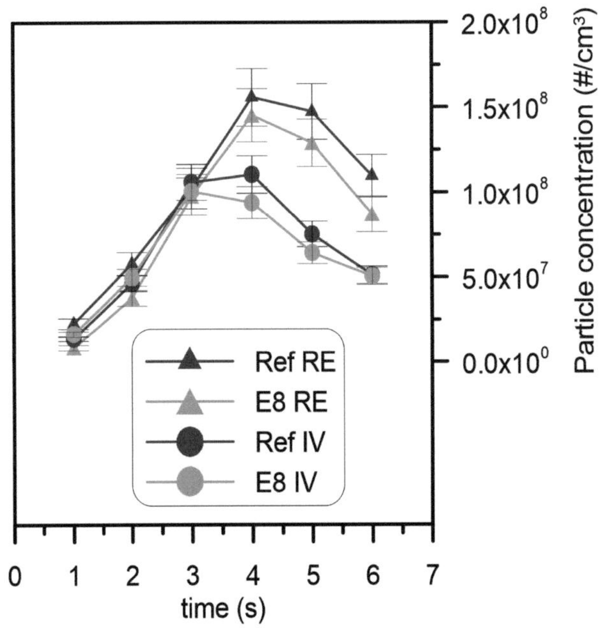

Figure 16.
Particle concentration profiles.

operation such as: relative fuel-air ratio (*Fr*), ambient temperature (T_{amb}, °C), exhaust gas temperature (T_g, °C), volumetric exhaust gas flow rate ($\dot{V}_g$, L/min), fuel mass flow rate ($\dot{m}_{fuel}$, g/s), and vehicle speed (v, km/h) for each operation sequence and fuel.

Figure 17 shows the relative fuel-air ratios, the smoke opacity and NOx concentration mean values registered during sequence C with both fuels. Important

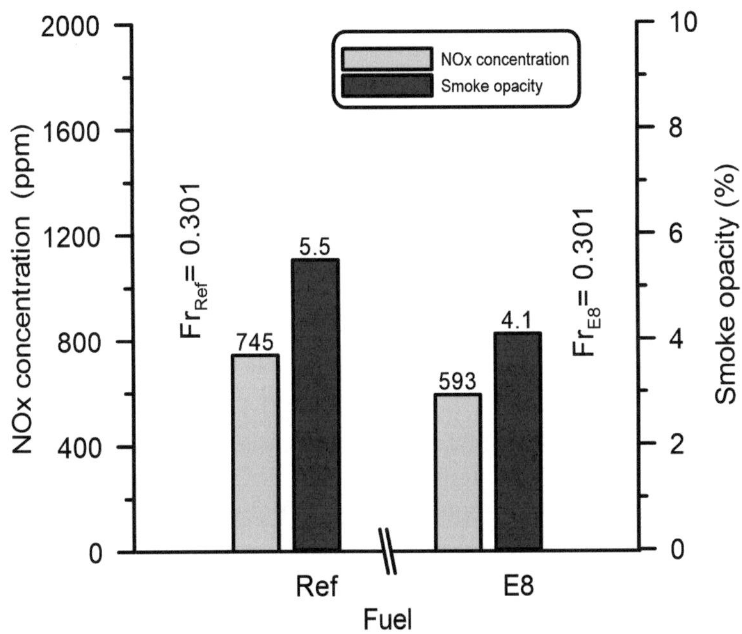

Figure 17.
Average values of relative fuel-air ratio, smoke opacity and NOx concentration from test fuels during the sequence C.

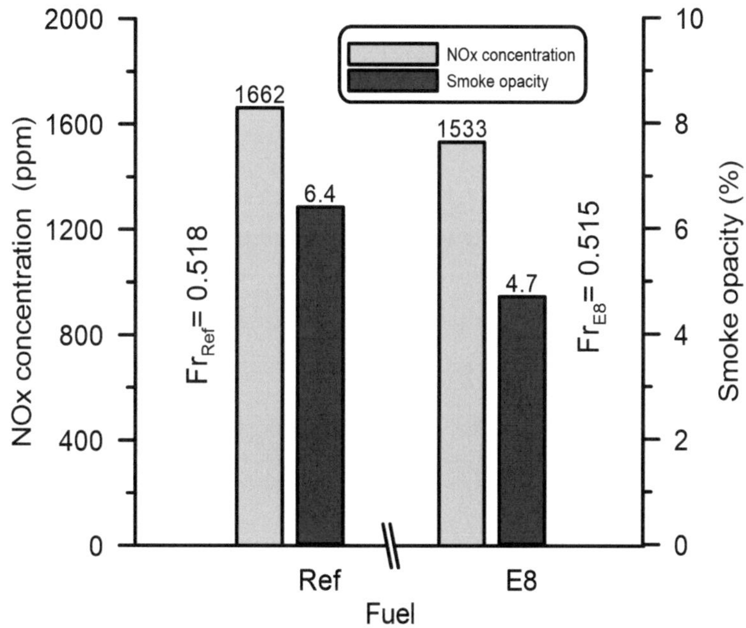

Figure 18.
Average values of relative fuel-air ratio, smoke opacity and NOx concentration from test fuels during the sequence W.

reductions of NOx concentration (20%) and smoke opacity (25%) was observed with E8 fuel. Smoke opacity reductions registered are comparable to results presented in [49].

Two factors contributed to provide reductions in NOx concentration higher than expected. First, the ambient temperature registered during the test with E8 fuel (9°C) lower than during the diesel tests. Second, machine velocity was 29% lower and thus the fuel consumption (as well as the exhaust gas flow) were lower (24%).

The average values of relative fuel-air ratios, smoke opacity and NOx concentration determined during the sequence W with both fuels are shown in **Figure 18**.

In this sequence NOx concentration and smoke opacity were reduced with E8 fuel blend 8% and 27% respectively. These decreases are in concordance with those presented in Ref.s [16, 47, 49]. During the test of this sequence, the ambient temperature was similar with both fuels. However, the E8 fuel consumption was 33% lower than the consumption of diesel fuel. This can be explained because the vibrating roller packed down harder ground when diesel fuel was used in comparison than when using E8. This difference, together with the lower machine velocity (16%) could partly explain the observed differences in opacity and NOx concentrations.

4. Strengths and weaknesses of the use of bioethanol mixed with diesel fuel

According to the results exposed in this chapter, most important strengths of the use of bioethanol blended with diesel fuel are the following:

- Molecular oxygen content of bioethanol is a key factor for reducing the emission of soot which is the main component of particulate matter emitted by diesel engines. Bioethanol molecular oxygen enables more complete combustion.

- Bioethanol, as linear chain alcohol, does not have aromatic compounds in its composition. In this sense, the addition of bioethanol to diesel fuel reduces the probability of soot nuclei formation in locally rich zones.

By the contrary, the use of bioethanol-diesel blends has several weaknesses:

- Limited miscibility reduces possibilities of using high percentages of bioethanol blended with diesel fuel. Low temperatures and high bioethanol concentration force the use of additives (such as biodiesel) to ensure the stability of bioethanol-diesel blends.

- Low lubricity is a key factor which negatively contributes on the lubrication of the injection systems and different engine parts. As occurred with miscibility, it is necessary the use of additives for improving lubricity properties of these blends.

- Bioethanol is highly hygroscopic. This forces to implement actions to avoid the water content increase of the blends. The water increase has double negative effect: decrease the miscibility of bioethanol-diesel blends and decrease their lubricity properties.

Other physicochemical properties of bioethanol can produce positive or negative effects on the engine operation when this alcohol is used blended with diesel fuel. However, the effects will depend on both the engine operating and ambient conditions. The high enthalpy of vaporization of bioethanol decreases the combustion process temperature, favoring lower NOx emissions under high engine load conditions but, under starting process or low engine load, this may provoke misfiring. Bioethanol has low cetane number that delay the start of combustion process which implies longer premixed phase of combustion. Also, bioethanol has low density and viscosity which can affect the fuel spray formation and its mixing with the inlet air. In this sense, the engine tuning should by adapted to fuel properties to maximize the benefits in terms of performance and emissions.

Author details

Octavio Armas Vergel[1*], Dolores Cárdenas[2], Reyes García-Contreras[1] and Carmen Mata[3]

1 Universidad de Castilla-La Mancha, Toledo, Spain

2 REPSOL, Madrid, Spain

3 Universidad de Castilla-La Mancha, Almadén, Spain

*Address all correspondence to: octavio.armas@uclm.es

References

[1] Agarwal AK. Biofuels (alcohols and biodiesel) applications as fuels for internal combustion engines. Progress in Energy and Combustion Science 2007; 33: 233–271. DOI: 10.1016/j.pecs.2006.08.003

[2] Ghadikolaei MA, Cheung Ch, Yung KF. Comparison between blended mode and fumigation mode on combustion, performance and emissions of a diesel engine fueled with ternary fuel (diesel-biodiesel-ethanol) based on engine speed. Journal of the Energy Institute 2019;92:1233-1250. DOI:10.1016/j.joei.2018.10.010

[3] Imran A, Varman M, Masjuki HH, Kalam MA. Review on alcohol fumigation on diesel engine: aviable alternative dual fuel technology for satisfactory engine performance and reduction of environment concerning emission, Renewable Sustainability Energy Review 2013;26:739-751. DOI: 10.1016/j.rser.2013.05.070

[4] Gomasta S, Mahla SK. An experimental investigation of ethanol blended diesel fuel on engine performance and emission of a diesel engine. International Journal of Emerging Technology 2012;3(1):74-79. ISSN (Online):2249-3255

[5] Can O, Çelikten I, Usta N. Effects of ethanol addition on performance and emissions of aturbocharged indirect injection Diesel engine running at different injection pressures. Energy Conversion and Management 2004;45: 2429-2440. DOI:10.1016/j.enconman.2003.11.024

[6] Abu-Qudais, M., Haddad,O., Qudaisat, M. The effect of alcohol fumigation on diesel engine performance and emissions. Energy Conversion & Management 2000;41: 389-399.DOI:10.1016/S0196-8904(99)00099-0

[7] Yao C, Cheung CS, Cheng C, Wang Y. Reduction of Smoke and NOx from Diesel Engines Using a Diesel/Methanol Compound Combustion System. Energy & Fuels 2007;21:686-691. DOI:10.1021/ef0602731

[8] Aakko P, Nylund NO, Westerholm M, Marjamäki M,Moisio M, Hillamo R, Mäkelä T. Emissions from Heavy-Duty Engine with and without After treatment Using Selected Biofuels. FISITA 2002 World Automotive Congress Proceedings, F02E195.

[9] Lapuerta M, Armas, O García-Contreras R. Effect of Ethanol on Blending Stability and Diesel Engine Emissions. Energy Fuels 2009;23:4343–4354. DOI:10.1021/ef900448m

[10] Pepiot-Desjardins P, Pistch H, Malhotra R, Kirby SR., Boehman AL. Structural group analysisfor soot reduction tendency of oxygenated fuels. Combustion and Flame 2008;154:191-205. DOI:10.1016/j.combustflame.2008.03.017

[11] Lapuerta M, Armas O, Herreros JM. Emissions from a diesel-bioethanol blends in anautomotive diesel engine. Fuel 2008;87:25-31. DOI:10.1016/j.fuel.2007.04.007

[12] Rakopoulos CD, Antonopoulos KA, Rakopoulos DC. Experimental heat release analysis and emissions of a HSDI diesel engine fuelled with ethanol-diesel fuel blends. Energy 2007;32: 1791-1808. DOI:10.1016/j.energy.2007.03.005

[13] Song C-L, Zhou Y-C, Huang R-J, Wang Y-Q. Influence of ethanol-diesel blended fuels on dieselexhaust emissions and mutagenic and genotoxic activities ofparticulate extracts. Journal ofHazardous Materials 2007;149: 355-363. DOI:10.1016/j.jhazmat.2007.03.088

[14] Giakoumis EG, Rakopoulos DC, Dimaratos AM, Rakopoulos D.C.

Exhaust emissions with ethanol or n-butanol diesel fuel blends during transient operation: Areview. Renewable and Sustainable Energy Reviews 2013;17:170-190. DOI:10.1016/j.rser.2012.09.017

[15] Ahmed I. Oxygenated Diesel: Emissions and Performance Characteristics of Ethanol-Diesel Blends in CI Engines. SAE 2001-01-2475.DOI: 10.4271/2001-01-2475

[16] Armas O, Cárdenas MD, Mata C. Smoke Opacity and NOx Emissions from a Bioethanol-Diesel Blend during Engine Transient Operation. SAE Technical Paper 2007-24-0131. DOI: 10.4271/2007-24-0131

[17] Armas O, García-Contreras R, Ramos Á. Pollutant emissions from engine starting with ethanol and butanol diesel blends. Fuel Processing Technology 2012;100:63-72. DOI: 10.1016/j.fuproc.2012.03.003

[18] Armas O, García-Contreras R, Ramos Á. Pollutant emissions from New European Driving Cycle with ethanol and butanol diesel blends. Fuel Processing Technology 2014;122:64–71. DOI:doi.org/10.1016/j.fuproc.2014.01.023

[19] Van Niekerk AS, Drew B, Larsen N, Kay PJ. Influence of blends of diesel and renewable fuels on compression ignition engine emissions over transient engine conditions. Applied Energy 2019;255: 113890. DOI:10.1016/j.apenergy.2019.113890

[20] Randazzo ML, Sodré JR. Exhaust emissions from a diesel powered vehicle fuelled by soybean biodiesel blends (B3–B20) with ethanol as an additive (B20E2–B20E5). Fuel 2011;90:98-103. DOI:10.1016/j.fuel.2010.09.010

[21] Mata C, Gómez A, Armas, O. The influence of ethanol-diesel blend on pollutant emissions from different bus fleets under acceleration transitions. Fuel 2017;209:322–328. DOI:10.1016/j.fuel.2017.08.018

[22] Kass M, Thomas J, Storey J, Domingo N, Wade J, Kenreck G. Emissions From a 5.9 Liter Diesel Engine Fueled With Ethanol Diesel Blends. SAE Technical Paper 2001-01-2018: DOI: 10.4271/2001-01-2018

[23] Armas O, Lapuerta O, Mata C, Pérez D. Online Emissions from a Vibrating Roller Using an Ethanol-Diesel Blend during a Railway Construction. Energy & Fuels 2009;23:2989–2996. DOI: 10.1021/ef900148c

[24] Merrit PM, Ulmet V, McCormick RL, Mitchell WE, Baumgard KJ. Regulated and Unregulated Exhaust Emissions Comparison for Three Tier II Non-Road diesel Engines Operating on Ethanol-Diesel Blends. SAE Paper 2005-01-2193. DOI:10.4271/2005-01-2193

[25] Hansen AC, Zhang Q, Lyne PWL. Ethanol–diesel fuel blends–a review. Bioresource Technology 2005;96:277–285. DOI:10.1016/j.biortech.2004.04.007

[26] Lapuerta, M., Armas, O., García-Contreras, R. Stability of diesel–bioethanol blends for use in diesel engines. Fuel 2007;86:1351–1357. DOI: 10.1016/j.fuel.2006.11.042

[27] Waterland LR, Venkates, S, Unnash S. Safety and Performance Assessment of Ethanol/Diesel Blends (E-Diesel). NREL/SR 2003;540:34817.

[28] Huang J, Wang Y, Li S, Roskilly AP, Yu H, Li H. Experimental investigation on the performance and emissions of a diesel engine fuelled with ethanol-diesel blends. Applied Thermal Engineering 2009;29:2484-2490. DOI:10.1016/j.applthermaleng.2008.12.016

[29] Di Y, Cheung CS, Huang Z. Experimental study on particulate

emission of a diesel engine fueled with blended ethanol-dodecanol-diesel. Journal of Aerosol Science 2009;40:101-112. DOI:doi.org/10.1016/j. jaerosci.2008.09.004

[30] Makareviciene V, Sendzikiene E, Janulis P. Solubility of multicomponent biodiesel fuel systems. Bioresource Technology 2005;96:611-616. DOI: 10.1016/j.biortech.2004.06.007

[31] Lei J, Shen L, Bi Y, Chen H. A novel emulsifier for ethanol–diesel blends and its effect on performance and emissions of diesel engine. Fuel 2012;93: 305-311. DOI:10.1016/j.fuel.2011.06.013

[32] McCormick RL, Parish R. Advanced Petroleum Based Fuels Program and Renewable Diesel Program. Milestone Report: Technical Barriers to the Use of Ethanol in Diesel Fuel. NREL/MP 2001;-540:32674.

[33] Kwachareon P, Luengnaruemitchai A, Jain-In S. Solubility of a diesel-biodiesel-ethanol blend, its fuel properties, and its emission characteristics from diesel engine. Fuel 2007;86:1053-1061. DOI:10.1016/j. fuel.2006.09.034

[34] Pires de Oliveira I, Lima Caires AR, Baskar K, Ponnusamyd S, Lakshmanan P, Veerappan V. Biodiesel as an additive for diesel-ethanol (diesohol) blend: physical-chemical parameters and origin of the fuels' miscibility. Fuel 2020;263:116753. DOI:10.1016/j. fuel.2019.116753

[35] Fernando S, Hanna M. Development of a Novel Biofuel Blend Using Ethanol-Biodiesel-Diesel Microemulsions: EB-Diesel. Energy & Fuels 2004; 18:1695-1703. DOI:10.1021/ef049865e

[36] Lapuerta M, García-Contreras R, Agudelo JR. Lubricity of Ethanol-Biodiesel-Diesel Fuel Blends. Energy & Fuels 2010;24:1374–1379. DOI:10.1021/ ef901082k

[37] Armas O, Mata C, Martínez-Martínez S. Effect of an ethanol–diesel blend on a common-rail injection system. International Journal of Engine Research 2012;13(5):417–428. DOI: 10.1177/1468087412438472

[38] McCormick RL and Parish R. Technical barriers to the use of ethanol in diesel fuel. National Renewable Energy Laboratory, Golden, Colorado, http://www.afdc.energy.gov/afdc/pdfs/ barriers_eDiesel.pdf (2001).

[39] Mickevičius T, Slavinskas S, Kreivaitis R. Effect of ethanol on performance and durability of a diesel common rail high pressure fuel pump. Transport 2016;31(6). DOI: 10.3846/ 16484142.2015.1058292

[40] Yahuza I, Dandakouta H. A Performance Review of Ethanol-Diesel Blended Fuel Samples in Compression-Ignition Engine. Journal of Chemical Engineering & Process Technology 2015;6(5). DOI: 10.4172/2157-7048.1000256

[41] Chandler K, Whalen M, Westhoven J. Final Results From The State Of Ohio Ethanol-Fueled Light-Duty Fleet Deployment Project. SAE Technical Paper 982531. DOI: 10.4271/982531

[42] Kane E, Huang Y, Mehta D, Frey CM, Tillerson Z, Chavis S. Refinement of a Dedicated E85 1999 Silverado with Emphasis on Cold Start and Cold Drivability. SAE Technical Paper 2001-01-0679. DOI: 10.4271/2001-01-0679

[43] Macian V, Bermúdez V, Payri R, Gimeno J. New technique for determination of internal geometry of a diesel nozzle with use of silicone methodology. Experimental Techniques 2003;27(2):39–43. DOI: 10.1111/j.1747-1567.2003.tb00107.x

[44] Armas O, Lapuerta M, Mata C. Methodology for the analysis of pollutant emissions from a city bus.

Measurement Science and Technology 2012;23:045302. DOI: 10.1088/0957-0233/23/4/045302

[45] Can Ö, Celikten I, Usta N. Effects of ethanol addition on performances and emissions of a turbocharged indirect injection Diesel engine running at different injection pressures. Energy Conversion and Management 2004;45: 2429-2440. DOI:10.1016/j.enconman.2003.11.024

[46] De-gang L, Huang Z, Lu X, Zhang W, Yang, J. Physico-chemical properties of ethanol-diesel blend fuel and its effect on performance and emissions of diesel engines. Renewable Energy 2005; 30:967-976. DOI: 10.1016/j.renene.2004.07.010

[47] Xing-cai L, Jian-guang Y, Wu-gao Z, Zhen H. Effect of cetane number improver on heat release rate emissions of high speed diesel engine fuelled with ethanol-diesel blend fuel. Fuel 2004;83: 2013-2020. DOI: 10.1016/j.fuel.2004.05.003.

[48] Mohanmmadi A, Ishiyama T, Kawanabe H, Horibe N. An optimal usage of recent combustion control Technologies for DI diesel engine operating on ethanol blended fuels. SAE paper 2004-01- 1886. DOI: 10.4271/2004-01-1866.

[49] Subramanian KA, Singal SK, Saxena M, Singhal S. Utilization of liquid biofuels in automotive diesel engine: An Indian perspective. Biomass and Bioenergy 2005;29:65-72. DOI: 10.1016/j.biombioe.2005.02.001

Chapter 7

Occupational Health Issue in a 2G Bioethanol Production Plant

Biancamaria Pietrangeli and Roberto Lauri

Abstract

The interest of scientists and health authorities in occupational risk related to biofuels production has recently increased due to the development of agro-industrial waste recycling processes in the framework of the European circular economy strategy and energy production from renewable sources. A common biofuel is the bioethanol, which is a leading candidate to substitute the gasoline as a transport fuel and it can be produced via biomass fermentation process. In biofuels production plants, some work activities in processing of biomass, are sources of airborne dust and the employers should demonstrate that adequate control measures have been implemented in order to prevent workers exposure. In the chapter, the production process of a 2G bioethanol plant has been analyzed in order to specify the process phases, which could generate occupational health issue related to airborne dust, and to provide technical recommendations.

Keywords: occupational health, bioethanol, biomass, airborne dust, bioaerosol

1. Introduction

In the European Union (EU), the original renewable energy directive 2009/28/EC establishes an overall policy for the production and promotion of energy from renewable sources. It requires to fulfill at least 20% of total energy needs by renewables within 2020. All EU countries must also ensure that at least 10% of their transport fuels comes from renewable sources within 2020 [1]. In December 2018, the revised renewable energy directive 2018/2001/EU came into force with a specific target for 2030: at least a 32% share of renewable energy consumption has to be achieved [2]. From an energy point of view, lignocellulosic biomass is increasingly recognized as a valuable resource, since it is an alternative to petroleum for the production of biofuels and chemicals. The first generation biofuels, such as agricultural bioethanol made from oilseed crops, have helped reduce greenhouse gases (GHG) emissions, but they also have a negative impact on water and soils, as well as competing for land used by food crops. By the year 2013, EU strategy was re-oriented towards "advanced" biofuels, made from waste or agricultural and forestry residues (second generation) or algae (third generation). In this context, bioethanol is one of the most important biofuels, which can be produced by fermentative processes of biomass, and therefore it is a leading candidate to substitute the gasoline as a transport fuel. A new approach involving the use of marginal land (i.e. land that is not suitable for food crop production or contaminated site) for the production of crops for biofuels industry is pursued in Italy and in many other countries, where the demand for high quality water resources, arable land, food and fossil fuels is rapidly growing. *Arundo donax* was selected as a potential crop for use in these areas, since it produces more cellulosic biomass and traps more contaminants, using less land and pesticides than any other alternative crops reported in the literature [3]. The direct job creation from advanced biofuels production is expected to grow in the future. Green jobs are activities characterized by previously evaluated risks, but with a different scope and exposition in connection with newly applied technology and therefore it is strategic and important to complete the risks assessment process with respect to new or emergent risks. An inclusive sustainability assessment of bioethanol production alternatives should incorporate the occupational health and safety (OH&S) assessment and it is necessary to integrate health and safety issues at an early stages of development of the industrial process in order to define tailored mitigation measures at full scale plant [4].

2. OH&S issue of bioethanol production from biomass

OH&S risks assessment in the biorefinery industry is a systematic examination of all aspects of the work undertaken to consider what could cause injury or harm, the possible hazards elimination, and, if not, which preventive or protective measures can be adopted to decrease the risk level [5]. In the biofuels production, some work activities, such as handling, storage and processing of biomass, are sources of organic dust (bioaerosol). The interest of scientific community and health authorities in bioaerosols has recently increased due to the development of waste recycling processes in the framework of circular economy, considering the wide range of adverse health outcomes associated with exposure to bioaerosol in workplaces. These include infections, immuno-allergic, non-allergic inflammatory and toxic effects [6–10]. Bioaerosol consist of live and dead micro-organisms either as individual micro-organisms or as aggregates, fragments and micro-organisms products, such as bacterial endotoxins, β (1–3)-D glucans and mycotoxins. All these biological agents can also be carried by other particles [11]. The viability of microorganisms is less important for effects, such as chronic bronchitis, asthma, toxic pneumonitis, hypersensitivity pneumonitis and lung function decline, as these effects can also depend on the exposure to non-viable microorganisms [12]. In agriculture, similar exposures to bioaerosol containing animal, plants, microbial components, can cause severe respiratory diseases, such as organic dust toxic syndrome or allergic alveolitis (e.g., farmers lung) [13–18]. Some technical surveys and occupational hygiene measurements at different biomass power plants showed that the occupationally harmful process steps were unloading, screening, crushing, conveying of fuels and the handling of biomass in silos. Unloading produced a great amount of organic dust, which spreads to the working stations. The main occupational exposure-associated health risks for workers were bacteria and fungi, which easily spread over the air during biomass processing [19]. The measured levels of exposure to bioaerosols were especially high during the unloading of peat and wood chips. Furthermore, biomass has also a tendency to decompose, creating exposure scenarios, that should be managed to minimize both microbial growth (e.g., spore formation, endotoxin release, etc.) and off-gassing of volatile organics or other gases (e.g., carbon monoxide). Besides the mechanical irritation caused by organic dust, the workers could also be exposed to chemical irritation caused by volatile organic compounds, such as terpenes emitted into the air in the gaseous phase during the outdoor storage of agriculture residues [20]. Hexanal from fatty acids oxidation is also emitted during the storage of solid wood fuels [21]. Multiple exposures to biological and chemical agents may simultaneously have synergistic health effects on workers lower and upper respiratory tracts [22]. Some Authors reviewed the available literature on OH&S issues associated with biomass-based power generation, considering the potential exposure scenarios and providing indications of hazards, which should be considered in the context of protecting the worker health through the development of monitoring and control plans [23]. A case-study facility for the production of second generation (2G) bioethanol has been considered in order to study some workers health issues. The study has been focused on occupational hazard related to workers exposure to airborne dust occurring during storage and processing of biomass, and on the preventive and protective measures aimed at controlling the exposure levels in the examined plant.

3. The case study: the bioethanol production plant

The case study has been focused on an Italian industrial plant (**Figure 1**), that produces bioethanol via fermentation of non-food biomass, based on a mix of available agricultural waste (bagasse of sugar cane, rice or wheat straw) and energy crops, such as *Arundo donax*, *Miscanthus spp, Panicum virgatum,* available from local supply chain (within a distance, which ranges between 40 km and 70 km). According to the provisions of the Directive 2009/28/EU [1], the production of 2G bioethanol from *Arundo donax* or from the residual of the corn and rice harvest is able to decrease the GHG emissions of over 80 percentage points compared to conventional processes for the production of petrol oil. On the contrary, the first generation bioethanol allows a reduction of only 22%. Furthermore, *Arundo donax* grows on marginal soils, requires a low consumption of water, fertilizers and territory, due to the high yield per hectare [24]. The 2G bioethanol plant, examined in the case study, allows the production of low cost sugars, which can be the platform of an industrial biorefinery aimed at producing a wide range of

Figure 1.
The bioethanol production plant.

intermediate chemicals from fine to bulk chemicals. The biorefinery was built on a decommissioned industrial site of about 15 hectares located in a rural area. The main quantitative data are:

- biomass storage hall capacity = 10,000 tons;
- bioethanol production capacity = 40,000 tons/year;
- electric energy production (the installed power is equal to 13 MW);
- water recycling = 100%.

Arundo donax (wet, about 60% humidity) and wheat straw (dry, about 10% humidity) are used to produce the biofuel. Arundo is shredded on the field and fed to the plant within a few days. As the straw is dry, it can be stored for longer periods without degrading and for this reason it can be used as an emergency biomass in case of Arundo lack. Biogas and lignin are the processing waste. The biogas is used to feed three boilers for the production of technological steam, while lignin is used to feed a larger boiler, which generates steam for the electric energy production, so that the plant operates in total autonomy.

The bioethanol production process is based on the following steps:

1. Pre-treatment. The pre-treatment is aimed at breaking down the structural components of the lignocellulosic matrix, separating the main polymeric components and making them accessible to the enzymatic hydrolysis in the following step. The pre-treatment minimizes the formation of biomass degradation products, such as furfural or hydroxymethylfurfural (5-HMF), which act as inhibitors of the fermentation process.

2. Enzymatic hydrolysis. The enzymatic hydrolysis is carried out by a mixture of cellulolytic enzymes (endocellulase, exocellulase and glycosidase), which allow to obtain hexoses, such as glucose, while the fraction composed by hemicellulose is split into a mixture of pentose sugars, in which xylose and arabinose are the most abundant components.

3. Fermentation. This step is performed by a selected strain of *Saccharomyces cerevisiae*, which is able to utilize all the monomeric sugars in order to achieve high yields of bioethanol.

4. Distillation. Bioethanol is separated from the residue of the fermentation broth in the distillation columns. The ethanol stream is successively subjected to a dehydration phase, which allows to obtain a 99% pure product.

4. Biomass storage and processing in the bioethanol facility

In the case study facility, biomass is stored in a covered shed with side movable gates. The storage area is about 20,000 square meters and is composed by two distinct areas:

1. zone for storage of the fresh biomass (e.g., *Arundo donax*, wood chips);

2. zone for storage of the dry biomass (straw).

In the storage hall, there are belt-conveyors, which move the biomass to the thermoelectric power plant and to the pre-treatment area. During biomass unloading, the gates are opened in order to limit the worker exposure to dust. In this working area, the cleaning practices are planned by the use of industrial sweepers (**Figure 2**) provided with dust control, which are suitable for use in areas, where explosive atmospheres could occur. Cleaning practices by compressed air are strictly forbidden. The cleaning operations are daily for the storage area floor and weekly for the equipment.

In the biomass hall, the daily work activities are the reception of the biomass (straw, arundo, wood chip) and its storage. The straw bales are unloaded and placed in a specifically signed area in one step by a forklift provided with the closed cabin (**Figure 3**). This procedure minimizes the bales moving and therefore the workers exposure to dust is strongly reduced. For the other biomasses, after unloading,

Figure 2.
Cleaning operation of the straw storage hall.

Figure 3.
Forklift used to transfer the straw bales.

the storage operations take place in automatic way. The worker operates, in remote control, cochleas and belt-conveyors, which transfer the biomass to three storage vessels and to the pre-treatment area.

In case of straw utilization, the straw bales are taken from the storage area and placed on the belt-conveyor of the grinding plant. The worker monitors the straw plant operation from the control room and therefore no manual operation is carried out by the employees. The specific worker task consists in ensuring the correct transport of the biomass from storage vessels and/or from straw plant to the boiler or to pre-treatment step. Inspections are planned along the walkways adjacent to the belt-conveyor. Considering that some areas of the storage hall are classified as Atex

Figure 4.
Straw feeding plant.

Figure 5.
Straw feeding plant: Shredder mill.

zones, the entire surface is protected by an automatic fire-fighting system, which is activated by temperature-sensitive strips placed on the ceiling and there are also wheeled fire extinguishers (their mass is equal to 30 kg) and an adequate number of portable fire extinguishers (their mass is equal to 6 kg). The biomass is successively transferred, through a second belt-conveyor, to a completely enclosed shredder mill aimed at suppressing the dust release. The mill cuts the biomass into small fragments, which are more suitable for the next fermentation step. The shredded biomass is successively moved to a silo by pneumatic transport system. Two magnets remove any small metal fragment, while a trap collects the heavy solid parts, such as stones, etc. The air is moved to a fabric filter, which traps the dust, before ejecting the air into the atmosphere through the chimney. The process activities are continuous and fully automated without the operator performing manual operations (**Figures 4** and 5).

5. Dust control strategy in the bioethanol production plant

In the bioethanol production plant, some process steps, such as unloading, storage and processing of the biomass, represent sources of risks for workers health, because they generate releases of airborne dusts in the work environment. In particular, the workers exposure to organic dust is associated with a wide range of health effects. Indeed, respiratory symptoms and lung function impairment are the most important health outcomes. In the facility, the plant design, the equipment and working methods have been implemented for limiting the workers exposure to airborne dust by containment/isolation principle. In addition, as part of dust control strategies, a dust monitoring program has been performed by the company in different working areas of the plant in order to assess the effectiveness of the adopted containment and control measures, but no specific measurement of the components of the organic dust has been carried out. The airborne dust sampling (twenty-two monitoring points) was conducted twice year (2016) in nine plant working areas, where the occupational exposure could be relevant:

1. the biomass storage areas (six monitoring points);
2. the biomass pre-treatment area (one point);
3. the power plant area (four points);
4. the lignin centrifugation area (two points);
5. the fermentation area (one point);
6. wood chipping process (two points);
7. the production control room (one point);
8. lab (one point).
9. offices areas (four points).

In **Figure 6**, the working areas of the airborne dust monitoring plan are reported. The dust sampling performed by the company in different working areas has been both static and personal. The area sampling provides a concentration, that reflects the general dust concentration in a defined area, while personal

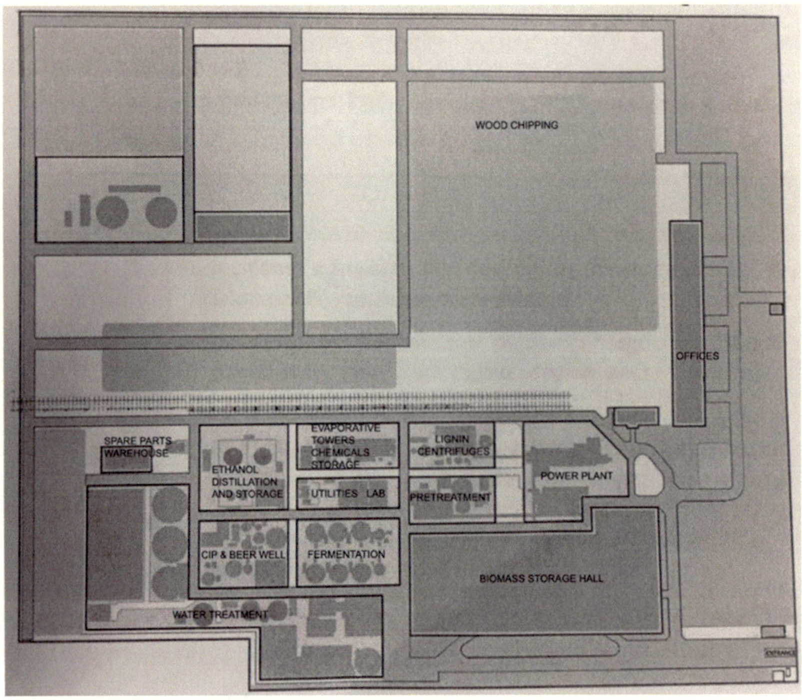

Figure 6.
The working areas of the airborne dust monitoring plan.

sampling provides a concentration measurement of airborne dust to which an individual is exposed. The airborne dust has been measured as the inhalable (or total dust) and respirable dust fraction, where the inhalable aerosol is the mass fraction of particles, which can be inhaled into the nose or mouth, and the respirable aerosol is the mass fraction of particles that may reach the alveoli in case of inhalation. Today, the American Conference of Governmental Industrial Hygienists (ACGIH) recommends TLVs guidance values equal to 10 mg/m^3 inhalable and 3 mg/m^3 respirable for insoluble or poorly soluble particles not otherwise specified (PNOS). The dust limits are based on personal exposure for a standard shift of eight consecutive hours and calculated as a time-weighted (TLV-TWA) average [25]. The inhalable and respirable dust fractions were determined by gravimetric methods of sampling and analysis, which are commonly used to measure quantities of airborne particulate matter collected from workplace atmospheres. The gravimetric dust sampler provides the time-weighted average concentration of dust. As the samplers determine the respirable dust, they are provided with a Dorr-Oliver cyclone, which separates respirable and oversize dust. The filters have been pre and post-weighted to determine the dust mass and to calculate the mean of dust concentration over sampling period [26, 27].

6. Airborne dust monitoring: results and discussion

Analysis of the results of airborne dust monitoring in twenty-two different working areas of the bioethanol production plant, shows that dust concentration, with regard to inhalable and respirable fractions, in all monitored stations is below the ACGIH guidance values. **Figures 7–9** show the results of airborne dust

Figure 7.
Biomass storage hall: airborne dust monitoring results (static sampling).

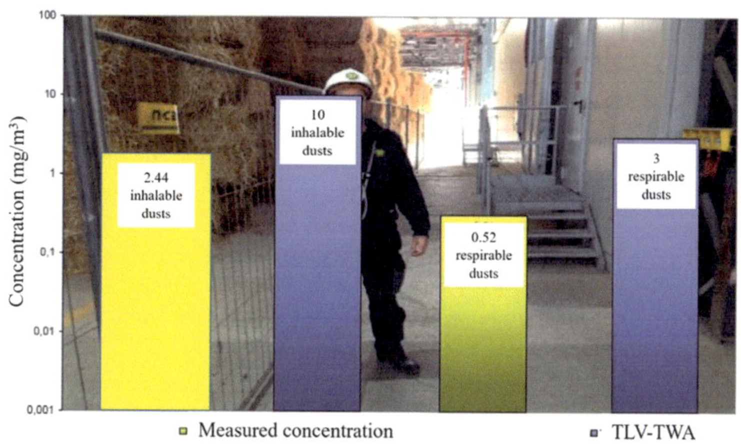

Figure 8.
Straw bales storage area: airborne dust monitoring results.

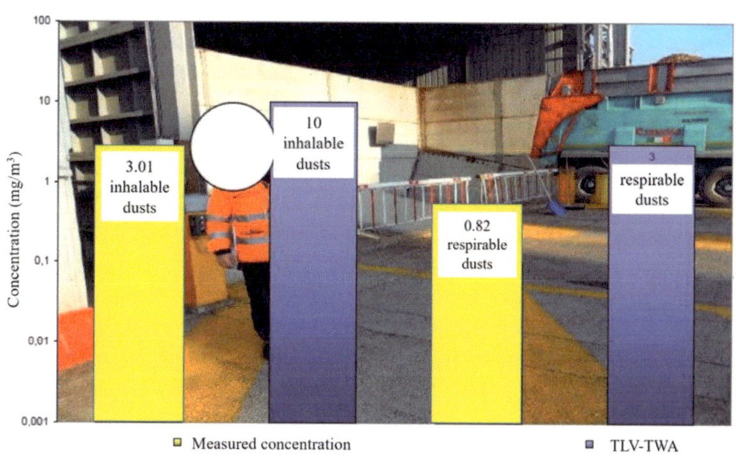

Figure 9.
Wood chips unloading area: airborne dust monitoring results (personal sampling).

concentrations in the working stations characterized by higher dust production. These areas respectively are the biomass storage hall, the straw bales storage area and wood chips unloading area. The concentrations have been compared with the ACGIH guidance values (TLV-TWA). The highest dust concentrations were found during the personal sampling (September 2016) carried out at the wood chips unloading area and were equal to 3.01 mg/m^3 for inhalable dust and 0.82 mg/m^3 for the respirable fraction (**Figure 9**). At the straw storage area, the personal sampling dust concentrations were equal to 2.44 mg/m^3 for inhalable dust and 0.52 mg/m^3 for the respirable fraction (**Figure 8**). In comparison, in the adjacent control room of the storage area, the exposure values were equal to 0.49 mg/m^3 for the inhalable fraction and 0.12 mg/m^3 for the respirable fraction. Another area exposed to dust production risk is that of the wood chipping process, which was characterized by measured exposure values equal to 2.59 mg/m^3 for inhalable dust and 1.98 mg/m^3 for respirable dust.

In order to provide a better assessment of workers exposure, the composition of the organic dust (bioaerosol) and its content of specific biological agents and/or their part or products, should be evaluated. In the examined plant, no specific measurement of the components of the organic dust has been carried out. The bioaerosol characterization is extremely complex and the prior knowledge of likely agents and their risk levels are required. Indeed, the bioaerosol may consists of live and dead microorganisms, either as individual microorganisms or as aggregates, fragments and microorganisms products, such as bacterial endotoxins, β (1–3)-D glucans and mycotoxins [28]. It follows that the measurement and interpretation of bioaerosol concentrations data are difficult. For example, the grain dust may contain fragments from grain, husk and straw, soil particles, pollen, bacterial spores and cells, fungal spores and hyphae, fragments and feces of mites and insects, microbial components such as endotoxins, glucans, peptidoglycans, mycotoxins, antigens, and allergens [29]. In agriculture, similar exposures to bioaerosol represent a major risk associated with severe respiratory diseases, such as organic dust toxic syndrome or allergic alveolitis (e.g., farmers lung) [13–18], but actually it is not clear which specific bioaerosol components primarily account for the observed health effects. In addition to these adverse health effects, some protective role of microbial exposure on atopy and atopic diseases has been suggested [6]. Certain microorganism-associated molecular patterns have been identified as agents, that might influence the development of the immune system, which in turn leads to protective effects for asthma and atopy [30].

A basic problem in quantitative assessment of exposure to bioaerosol is the variability of microbial agents, which can be substantially greater than that commonly found for chemical agents, because microorganisms may rapidly proliferate in case of favorable conditions. Rohr et al. [23] have shown that the tests results (by culture based-methods) of fungal and bacterial levels in the bioaerosol of biomass-based power stations indicated extremely variable concentrations. The workers exposure to inhalable airborne fungi, bacteria, endotoxin at five biofuel heating or power plants showed that the exposure levels differed among the plants. This was due to the different process equipment, tasks and the handled biofuels [31]. In particular, it should be noted that bioaerosol analysis by culture based-methods could underestimate the real contamination of the workplace. Furthermore, the viability of microorganisms is less important for health effects, which can be also caused by exposure to non-viable microorganisms [12]. Nowadays, there are several gaps in knowledge concerning each step of the biological risk assessment, with regard to hazard identification, exposure assessment, and, above all, relationship between exposure and health risk. A systematic review of the studies on health effects of bioaerosol concluded that none of the analyzed studies provided suitable dose–response relationships for derivation of exposure limits [32]. The main reasons were:

1. lack of studies with valid dose–response data;

2. diversity of employed measurement methods for microorganisms and bioaerosol emitted by facilities;

3. heterogeneity of health effects;

4. insufficient exposure assessment.

Indeed, it is important to highlight that health effects of exposure to bioaerosol can substantially vary from person to person, because the human response to exposure to biological agents depends on individual susceptibility to infections and allergies [33]. A variable human response has been described for workers exposure to organic dust in different workplace areas and it was shown that the dust composition may play an important role in determining its health effects [34]. Because of lack of health-related exposure limits for bioaerosol components based on toxicological or epidemiological studies from the workplaces or environmental health [32, 35], few occupational exposure standards (not OELs) have been set by regulatory organizations, such as the ACGIH or the AIHA [10, 28, 36]. Although the research in this field is going on, setting OELs requires more exposure–response data derived from a greater number of animal models and, in particular, epidemiological studies of human exposure. Standardized and reproducible measurement methods are also required to compare studies in different environments [10, 35]. Considering that, it is not very likely that OELs for biological agents in bioaerosol will be developed in short times [35], the TLV referred to "particulates not otherwise regulated" [25] is used in lack of more specific values. Besides the availability of health-related exposure limits (OELs) for biological agents and additional studies on the respiratory health of biofuels plant workers, the identification of exposure indicators, easy to monitor, such as airborne dusts, can be useful tools for the routine assessment of workers exposure. Furthermore, some authors indicated that inhalable dust, at least in some workplaces, showed a good correlation with total bacterial counts and bacterial endotoxins and therefore it could be proposed as a valid indicator of human exposure to bioaerosol in workplaces with similar exposures [37, 38]. Considering that the scientific evidence on health effects of bioaerosol emissions related to biomass processing in bioethanol production plants is still limited, all valuable preventive technical measures, in accordance with the controls hierarchy [39, 40], should be taken into account for decreasing the exposure to airborne dust.

The case study facility has adequate plant layout and control measures aimed at limiting the workers exposure to organic dust. The preventive technical measures include:

- Separating the dusty operations from non-dusty activities;

- Reducing the speed of all vehicles near the plant;

- Using enclosed dusty machines (e.g., mill, which shreds the biomass) in order to reduce the spreading of bioaerosol in the work environment;

- Automated production and remotely controlled operations without the presence of workers;

- Reducing the discharge points and amount of materials, which have to be transferred;

- Employing frequent cleaning operations in the biomass storage area by good practices (industrial sweepers provided with dust control or mobile vacuum cleaners used to clean up possible spillages along the belt-conveyers).

Within the dust control strategy, the monitoring program ensures a reasonable representation of exposure to airborne dust for specific work activities (personal workers sampling) and for the working areas (stationary sampling) in the bioethanol production plant. In addition, the dust monitoring is a valid tool in order to assess the effectiveness of airborne dust containment measures. With regard to the experimental evidence that the inhalable dust could be a suitable tool for assessing the workers exposure to bioaerosol, by simple and not expensive methods [37, 38], it would be advisable planning specific studies in order to verify these observations in biofuels production plants. Furthermore, tailored workers health surveillance studies should be performed in order to link the bioaerosol exposure to the respiratory health of biofuels plant workers.

7. Conclusions

In bioethanol production plants, some work activities in the processing of biomass are sources of airborne organic dust. In the case study facility, the plant design, the equipment and working methods have been implemented for limiting the workers exposure to airborne dust by containment/isolation principle. In addition, a dust monitoring program has been performed in order to assess the effectiveness of the adopted containment and control measures, but no specific measurement of the components of the bioaerosol has been carried out. In order to overcome the current knowledge gaps in establishing agreed bioaerosol monitoring protocols and developing reliable dose–response data, the potential risk should be managed by a precautionary approach, such as in other comparable industries [41]. Every worker, even if only potentially exposed, must be protected using the best practices based on the most up-to-date scientific knowledge and on the current level of technological development. In order to prevent respiratory impairment among workers of bioethanol production plants, the employers should demonstrate that adequate control measures have been developed in order to keep the exposure to dust as low as possible. In the examined plant, the preventive technical measures, the work equipment and working methods of biomass processing ensure the observation of the ACGIH guidance values for inhalable and respirable dusts in the workplace. Limited information is available on workers health surveillance programs in the biofuels production plants and therefore there is the real need of data collection on workers symptoms and diseases associated with the exposure to airborne dust in order to improve the knowledge on health outcomes of highest concern, such as respiratory impairment, airways irritation and sensitization.

Author details

Biancamaria Pietrangeli* and Roberto Lauri
INAIL Research, Department of Technological Innovations and Safety of Plants, Products and Human Settlements, Rome, Italy

*Address all correspondence to: b.pietrangeli@inail.it

References

[1] European Union. Directive 2009/28/EC of the European Parliament and the Council of 23 April 2009 on the promotion of the use of energy from renewable sources and amending and subsequently repealing Directives 2001/77/EC and 2003/30/EC. Official Journal of the European Union. L 140/16, 5.6.2009. pp. 16-62.

[2] European Union. Directive 2018/2001/EC of the European Parliament and the Council of 11 December 2018 on the promotion of the use of energy from renewable sources. Official Journal of the European Union. L 328/82, 21.12.2018. pp. 82-209.

[3] Accardi DS, Russo P, Lauri R, Pietrangeli B, Di Palma L. From soil remediation to biofuel: process simulation of bioethanol production from *Arundo donax*. Chemical Engineering Transactions. 2015; 43: 2167-2172. DOI: 10.3303/CET1543362.

[4] Lauri R, Pietrangeli B. Biorefinery Safety: A Case Study Focused on Bioethanol Production. In: Beschkov V, editor. Biorefinery Concepts, Energy and Products. London: IntechOpen; 2019. ISBN 978-1-78985-676-7.

[5] Accardi D S, Bubbico R, Di Palma L, Pietrangeli B. Environmental and safety aspects of integrated biorefineries (IBR) in Italy. Chemical Engineering Transactions. 2013; 32:169-174. DOI: 10.3303/CET1332029.

[6] Douwes J, Thorne P, Pearce N, Heederik D. Bioaerosol health effects and exposure assessment: progress and prospects. Annals of Occupational Hygiene. 2003; 47: 187-200. DOI: 10.1093/annhyg/meg032.

[7] Dutkiewicz J. Bacteria and fungi in organic dust as potential health hazard. Annals of Agricultural and Environmental Medicine. 1997; 4: 11-16.

[8] Smit LA, Heederik D, Doekes G, Blom C, van Zweden I, Wouters I. Exposure-response analysis of allergy and respiratory symptoms in endotoxin-exposed adults. European Respiratory Journal. 2008; 31: 1241-1248. DOI: 10.1183/09031936.00090607.

[9] Bunger J, Antlauf-Lammers M, Schulz TG, Westphal GA, Muller MM, Ruhnau P, Hallier E. Health complaints and immunological markers of exposure to bioaerosols among biowaste collectors and compost workers. Occupational Environmental Medicine. 2000; 57 (7): 458-464. DOI: 10.1136/oem.57.7.458.

[10] Eduard W, Heederik D, Duchaine C, Green BJ. Bioaerosol exposure assessment in the workplace: the past, present and recent advances. Journal of Environmental Monitoring. 2012; 14: 334-339. DOI: 10.1039/c2em10717a.

[11] American Conference of Governmental Industrial Hygienists (ACGIH). Bioaerosols: Assessment and Control. Cincinnati, OH: ACGIH Press; 1999. ISBN:978-1-882417-29-1.

[12] Thorne PS, Duchaine C, Douwes J, Eduard W, Gorny R, Jacobs R, Reponen T, Schierl R, Szponar B. Working Group Report 4: Exposure assessment for biological agents. American Journal of Industrial Medicine. 2004; 46 (4): 419-422. DOI: 10.1002/ajim.20066.

[13] Pinheiro GA, Syamlal G, Antao V. Twenty-three years of hypersensitivity pneumonitis mortality surveillance in the United States. American Journal of Industrial Medicine. 2006; 49 (12): 997-1004. DOI: 10.1002/ajim.20405.

[14] Lacey J, Crook B. Fungal and actinomycete spores as pollutants of the workplace and occupational allergens. Annals of Occupational Hygiene. 1988; 32 (4): 515-533. DOI. 10.1093/annhyg/32.4.515.

[15] Burrell R. Immunotoxic reactions in the agricultural environment. Annals of Agricultural and Environmental Medicine. 1995; 2 (1): 11-20.

[16] Sorenson WG, Shahan TA, Lewis DM. Activation of alveolar macrophages by conidia of common fungi associated with organic dust toxic syndrome. In: Samson RA, Flannigan B (Eds). International Workshop: Health Implications of Fungi in Indoor Environments. Amsterdam: Elsevier; 1994. p.325-343.

[17] Von Essen S, Fryek, J, Nowakowski B, Wampler M. Respiratory symptoms and farming practices in farmers associated with an acute febrile illness after organic dust exposure. Chest. 1999; 116: 1452-1458. DOI: 10.1378/chest.116.5.1452.

[18] Olenchock SA. Health effects of biological agents: the role of endotoxins. Applied Occupational and Environmental Hygiene. 1994; 9(1): 62-64. DOI: 10.1080/1047322X.

[19] Laitinen S, Laitinen J, Fagernas L, Korpijarvi K, Korpinen L, Ojanen K, Aatamila M, Jumpponen M, Koponen H, Jokiniemi J. Exposure to biological and chemical agents at biomass power plants. Biomass & Bioenergy. 2016; 93: 78-86. DOI: 10.1016/j.biombioe.2016.06.025.

[20] Rupar K, Sanati M. The release of terpenes during storage of biomass. Biomass & Bioenergy. 2005; 28 (1): 29-34. DOI: 10.1016/j.biombioe.2004.06.005.

[21] Granström KM. Emissions of hexanal and terpenes during storage of solid wood fuels. Forest Products Journal. 2010; 60 (1): 27-32. DOI: 10.13073/0015-7473-60.1.27.

[22] European Agency for Safety and Health at Work. Expert forecast on Emerging Biological Risks related to Occupational Safety and Health. European Risk Observatory Report. EU-OSHA. Luxembourg: Publications Office of the European Union, 2007. ISBN: 1830-5946.

[23] Rohr AC, Campleman SL, Long CM, Peterson MK, Weatherstone S, Quick W, Lewis A. Potential occupational exposures and health risks associated with biomass-based power generation. International Journal of Environmental Research and Public Health. 2015; 12(7): 8542-8605. DOI: 10.3390/ijerph120708542.

[24] Accardi D S, Lauri R, Pietrangeli B. Comparative study in the safety aspects of the production of ethanol from renewable and non-renewable sources. Procedia Environmental Science, Engineering and Management. 2015; 2: 161-168.

[25] American Conference of Governmental Industrial Hygienists (ACGIH). TLVs and BEIs based on the documentation of the threshold limit values for chemical substances and physical agents & biological exposure indices. Cincinnati, OH: ACGIH Press; 2012. ISBN:978-1-607260-48-6.

[26] National Institute for Occupational Safety and Health (NIOSH). Particulates not otherwise regulated, total: Method 0500, Issue 2: 15 August 1994. NIOSH Manual of Analytical Methods (NMAM), 4th Edition.

[27] National Institute for Occupational Safety and Health (NIOSH). Particulates not otherwise regulated, respirable: Method 0600, Issue 3: 15 January 1998. NIOSH Manual of Analytical Methods (NMAM), 4th Edition.

[28] National Institute for Occupational Safety and Health (NIOSH). Sampling and characterization of bioaerosols. In: Lindsley DG, Green BJ, Blachere FM, Martin SB, Law BF, Jensen PA, Schafer MP (Eds.). NIOSH Manual of Analytical

Methods (NMAM), 5th Edition. Washington DC: Chapter BA, CDC Press; 2017. p.115-214.

[29] Eduard W, Straumforms Halstensen A. Quantitative exposure assessment of organic dust. Scandinavian Journal of Work, Environment & Health. Suppl. 2009; (7): 30-35.

[30] Heederik D, von Mutius E. Does diversity of environmental microbial exposure matter for the occurrence of allergy and asthma? Journal Allergy Clinical Immunology. 2012; 130:44-50. DOI:10.1016/j.jaci.2012.01.067.

[31] Madsen AM. Exposure to Airborne Microbial Components in Autumn and Spring During Work at Danish Biofuel Plants. Annals of Occupational Hygiene. 2006; 50 (8): 821-831. DOI: 10.1093/annhyg/mel052.

[32] Walser SM, Gerstnera DG, Brennera B, Büngerb J, Eikmannc T, Janssena B, Kolba S, Kolkd A, Nowake D, Raulfb M, Sagunskif H, Sedlmaierg N, Suchenwirth R, Wiesmülleri G, Wollinh KM, Tesserauxj I, Herra C. Evaluation of exposure–response relationships for health effects of microbial bioaerosols - A systematic review. International Journal of Hygiene and Environmental Health. 2015; 218 (7): 577-589. DOI: 10.1016/j.ijheh.2015.07.004.

[33] Segal S, Hill AVS. Genetic susceptibility to infectious disease. Trends in Microbiology. 2003; 11(9):445-448. DOI. 10.1016/s0966-842x(03)00207-5.

[34] Smith LM, Weissenburger-Moser LA, Heires AJ, Bailey KL, Romberger DJ, LeVan TD. Epistatic effect of TLR-1, −6 and −10 polymorphisms on organic dust-mediated cytokine response. Genes and Immunity. 2017; 18: 67-74. DOI:10.1038/gene.2016.51.

[35] European Agency for Safety and Health at Work. Biological agents and work-related diseases: results of a literature review, expert survey and analysis of monitoring systems. EU-OSHA. Luxembourg: Publications Office of the European Union; 2019. ISBN: 978-92-9479-142-9.

[36] Schlosser O. Bioaerosols and health: current knowledge and gaps in the field of waste management. Detritus. 2019; 5: 11-125. DOI: 10.31025/2611-4135/2019.13786.

[37] Smit LA, Wouters I, Hobo MM, Eduard W, Doekes G, Heederik D. Agricultural seed dust as a potential cause of organic dust toxic syndrome. Occupational and Environmental Medicine. 2006; 63:59-67. DOI: 10.1136/oem.2005.021527.

[38] Schlosser O, Robert S, Debeaupuis C, Huyard A. Inhalable dust as a marker of exposure to airborne biological agents in composting facilities. Waste Management. 2018; 81: 78-87. DOI: 10.1016/j.wasman.2018.09.051.

[39] Occupational Health and Safety Assessment Series (OHSAS) 18001:2007. Occupational health and safety management systems - Requirements. OHSAS Project Group 2007. ISBN 978 0 580 50802 8.

[40] Occupational Safety and Health Administration. Recommended Practices for Safety and Health Programs. ISO 45001:2018. OSHA Ed., Washington, DC. 2016.

[41] Pietrangeli B, Lauri R. Biogas production plants. A methodological approach for occupational health and safety improvement. In: Nageswara-Rao M, Soneji J, editors. Advances in Biofuel and Bioenergy. London: IntechOpen; 2018. ISBN 978-1-78923-286-8.